Foun**dation**

Math**ematics and Statistics**

Foundation
Mathematics and Statistics

Thomas Bending

Australia · Canada · Mexico · Singapore · Spain · United Kingdom · United States

THOMSON

Foundation Mathematics and Statistics
Thomas Bending

Middlesex
University
PRESS

Series Editor		**Publishing Partner**
Walaa Bakry, Middlesex University		Middlesex University Press

Publishing Director	**Commissioning Editor**	**Managing Editor**
John Yates	Gaynor Redvers-Mutton	Celia Cozens
Production Editor	**Manufacturing Manager**	**Marketing Manager**
Amy Blackburn	Helen Mason	Mark Lord
Production Controller	**Text Design**	**Cover Design**
Maeve Healy	Design Deluxe, Bath	Matthew Ollive

Typesetter	**Printer**
Pages Unlimited/Keyline Consultancy, Newark	C&C Offset Printing Co., Ltd, China

Contents

The FastTrack Series

Thomson Learning and Middlesex University Press have collaborated to produce a unique collection of textbooks which cover core, mainstream topics in an undergraduate computing curriculum. FastTrack titles are instructional, syllabus-driven books of high quality and utility. They are:

- **For students**: concise and relevant and written so that you should be able to get 100% value out of 100% of the book at an affordable price
- **For instructors**: classroom tested, written to a tried and trusted pedagogy and market-assessed for mainstream and global syllabus offerings so as to provide you with confidence in the applicability of these books. The resources associated with each title are designed to make delivery of courses straightforward and linked to the text.

FastTrack books can be used for self-study or as directed reading by a tutor. They contain the essential reading necessary to complete a full understanding of the topic. They are augmented by resources and activities, some of which will be delivered online as indicated in the text.

How the series evolved

Rapid growth in communications technology means that learning can become a global activity. In collaboration, Global Campus, Middlesex University and Thomson Learning have produced materials to suit a diverse and innovating discipline and student cohort.

Global Campus at the School of Computing Science, Middlesex University, combines local support and tutors with CD-ROM-based materials and the Internet to enable students and lecturers to work together across the world.

Middlesex University Press is a publishing house committed to providing high-quality, innovative, learning solutions to organisations and individuals. The Press aims to provide leading-edge 'blended learning' solutions to meet the needs of its clients and customers. Partnership working is a major feature of the Press's activities.

Together with Middlesex University Press and Middlesex University's Centre for Learning Development, Global Campus developed FastTrack books using a sound and consistent pedagogic approach. The SCATE pedagogy is a learning framework that builds up as follows:

- **Scope:** Context and the learning outcomes
- **Content:** The bulk of the course: text, illustrations and examples
- **Activity:** Elements which will help students further understand the facts and concepts presented to them in the previous section. This promotes students' active participation in their learning and in creating their understanding of the unit content
- **Thinking:** These elements give students the opportunity to reflect and share with their peers their experience of studying each unit. There are *review questions* so that the students can assess their own understanding and progress
- **Extra:** Further online study material and hyperlinks which may be supplemental, remedial or advanced.

Foundation mathematics and statistics

Foundation Mathematics and Statistics ensures that you can demonstrate competency in a range of mathematical tools required for technical subjects, and gives you the confidence you'll need in the classroom. Explanations of mathematical tools are supported by real-world examples and graded exercises to enable you to practise and revise each topic.

The book starts from the basics of arithmetic and algebraic manipulation, and covers linear equations, both from an algebraic viewpoint and by considering graphs. It goes on to cover polynomials, exponentials and logarithms, again presented both algebraically and graphically. Particular attention is paid to the conversion of a textual problem into an algebraic form, so you can use the methods to tackle real-world problems. The book gives a general grounding in proportions, ratios and percentages, together with an introduction to set theory and a discussion of probability. The summary and presentation of statistical data and the drawing of histograms are also covered.

Using this book

There are several devices which will help you in your studies and use of this book. **Activities** usually require you to try out aspects of the material which have just been explained, or invite you to consider something which is about to be discussed. In some cases, a response is provided as part of the text that follows – so it is important to work on the activity before you proceed! Usually, however, a formal answer will be provided in the final section of each chapter.

The **time bar** indicates *approximately* how long each activity will take:

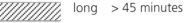 short   < 10 minutes

medium 10-45 minutes

long   > 45 minutes

 Review questions are (usually) short questions at the end of each chapter to check you have remembered the main points of a chapter. They are a useful practical summary of the content, and can be used as a form of revision aid to ensure that you remain competent in each of the areas covered.

About the author

Thomas Bending

Thomas Bending is head of the mathematics and statistics group at Middlesex University, where he has taught mathematics and statistics for business and computing since 1996. He is a combinatorial mathematician whose research interests include finite designs and geometries and the use of computing in mathematics teaching. He has developed this book from a version used at Middlesex University which was co-written with John Hammond, Gary Hearne, Matthew Jones, Alison Megeney and Catherine Minett-Smith.

Visit the accompanying website at **www.thomsonlearning.co.uk/fasttrack** and click through to the appropriate booksite to find further teaching and learning material including:

For Students

- Activities
- Multiple choice questions for each chapter.

For Lecturers

- Downloadable PowerPoint slides.

CHAPTER 1

Introduction to algebra

OVERVIEW

This chapter introduces you to the basic ideas and techniques of algebra. It builds on your current knowledge of the subject, including the notion of numbers, the manipulation of numbers in arithmetic; you will be assumed to know how to perform the usual operations with numbers, such as addition, subtraction, multiplication and division, as well as powers and roots. You will be assumed, too, to have some familiarity with algebra. The aim of this chapter is to enrich and deepen this familiarity. These basic ideas and techniques will help you in the later chapters, as well as in the world at large.

Learning outcomes	On completion of this chapter, you should be able to:

- Recognise and use the language of algebra

- Use and interpret indices

- Evaluate various kinds of algebraic expression

- Manipulate various kinds of algebraic expression.

1.1 Introduction

In this chapter you will be introduced to some of the essential ideas and language of algebra, the most important of which is the notion of *universalising*. You will have a chance to revise the operations of arithmetic, but this time, in a more general context.

We begin with an overview of mathematics. Then you are introduced to the language of algebra and its symbolism. This symbolism is explored with various techniques, including the use of letters of the alphabet to represent numbers, evaluation of an expression by substitution, some forms of simplification of algebraic expressions and other manipulations of algebraic expressions. You are reminded of the importance of the convention setting out the order in which algebraic operations must be performed.

What is mathematics?

Mathematics starts with counting and measuring.

We split the subject of mathematics into sub-topics:

- **Arithmetic** provides the rules for combining numbers and measurements to obtain results which are useful in the real world
- **Algebra and calculus** provide a language for expressing those rules in a general form which can be widely applied
- **Geometry and trigonometry** provide rules for combining the measurements of objects whose relative positions in space are important. They borrow language structures from algebra and calculus.

Who does mathematics?

The young child who is absorbed in exploring his/her surroundings has begun to experience the power of being able to predict, and therefore control, the effect of changes he/she makes. If he/she lets go of a toy, it falls to the floor, again and again. Older children begin to observe the patterns which capture the rules of prediction. They have begun to experience mathematics as a science. Scientists, engineers and business people use mathematics.

It is the beauty of these patterns which inspire those who may not be particularly interested in controlling the natural world. The patterns may subconsciously help them to enjoy art, architecture, music and the decorations employed in all kinds of craft. They are experiencing mathematics as an art.

Pure mathematicians enjoy exploring the patterns and structures for their own sake.

The advantages of knowing mathematics

'Mathematics gets your thoughts into parts that other disciplines can't reach.' (Prof JCR Hunt, President of the Institute of Mathematics and its Applications, and Chief Executive of the Meteorological Office 1995)

1.2 Order of operations

Consider the expression 7+4-1.

If you do the addition first, you get 7+4-1=11-1=10.

On the other hand, if you do the subtraction first, you get 7+4-1=7+3=10.

That is, you get the same answer 10 whatever order you do the two operations in.

Now consider the expression 2+3x4.

If you do the addition first, you get 2+3x4=5x4=20.

On the other hand, if you do the multiplication first, you get 2+3x4=2+12=14.

To avoid such confusions, mathematicians have agreed a certain *convention* about the order in which mathematical operations are performed.

If there are no brackets, then multiplication and division (and powers of numbers, which is a form of multiplication) are always performed before addition and subtraction.

If there are brackets in the expression, then these are performed before anything else.

This convention is summarised by the acronym **BoDMAS**:

> **B**rackets
>
> **D**ivide
>
> **M**ultiply
>
> **A**dd
>
> **S**ubtract.

It is usual for calculators and computers to have this 'BoDMAS' convention programmed into them so that they automatically carry out the operations in the correct order. Try with your own calculator.

1.3 Negative numbers

A negative quantity is like a debt or a time before zero hour in the countdown to some event (like blastoff).

Debts can be added together to make bigger debts, so:

> -4+(-6)=-10.

Debts can be multiplied to make bigger debts, so:

> 3(-4)=-12.

Debts can be partly or wholly cancelled by some credit, so:

> -9+7=-2

but

> -9+11=2.

Subtracting a debt from one's bank account is actually the addition of some credit, so:

> (-5)=+5.

Subtracting two debts of £5 from one's bank account is actually the addition of two lots of credit, so:

-2(-5)=+10

Rules for multiplying and dividing

Like signs give a plus; unlike signs give a minus.

These rules can be summarised nicely in two tables:

x	+	-		÷	+	-
+	+	-		+	+	-
-	-	+		-	-	+

This shows: (+)x(+)=(+)   This shows: (+)÷(+)=(+)

(+)x(-)=(-)   (+)÷(-)=(-)

(-)x(+)=(-)   (-)÷(+)=(-)

(-)x(-)=(+)   (-)÷(-)=(+)

1.4 Algebraic language

Mathematical shorthand – the use of symbols

Single, usually lower-case, letters are used to represent numbers.

The multiplication sign is omitted because $2a$, for example, means 2 times a, and ab means a times b.

The language provides a shorthand for expressing a formula. For example, $v=lbh$ is the formula for the volume, v, in cc of a box (cuboid) with length l cm, breadth b cm and height h cm.

Algebra is, in fact, the great universaliser. If we want to find the volume of a particular box with length 5cm, breadth 3cm and height 2cm, we **substitute** the values 5, 3 and 2 for l, b and h respectively to obtain the result 5 * 3 * 2 to obtain the volume 30cc (*not* 532cc). Note that we more usually write this calculation as 5x3x2.

Notice that the symbol * is used for multiplication in much printed material and in computer packages. Normally, though, we use x or • (But be careful to distinguish these from the letters x or X and from the decimal point.)

Letters used to represent numbers are called **variables** because they are not fixed until we substitute the values we want to use with the formula.

In this case, v is the **subject of the formula** and is the **unknown** value we want to find. The skill of changing the subject of a formula is addressed in the section towards the end of this chapter, *Changing the subject of a formula* (also known as transposition).

Observe that in the algebra of 'ordinary numbers' the order of multiplication does not matter. This means, for example, that 2x3=3x2, or $axb=bxa$. (The same is true of addition so that $a+b=b+a$.) This law, known as the **commutative law** (of multiplication or of addition), always applies to ordinary numbers. But should you go on to study matrices, for example, you will see that you can have perfectly good (= useful) algebra where this rule, that we take so much for granted, does not hold (in the case of matrix multiplication).

Another piece of shorthand we use in mathematics is an expression like a^2 where the 2 sits above the line. This is called a **power of** a and the 2 is called an **index**. a^2 is a shorthand for $a \times a$ (or, equally, $a \bullet a$). Thus, 3^2 means 3×3 which, of course, has the value 9. Thus, 3^2, 3×3 and 9 are three ways of expressing the number 9.

In a similar way, a^3 is shorthand for $a \times a \times a$, a^4 is shorthand for $a \times a \times a \times a$; and so on. You will see a lot more on indices in the last section of this chapter, *Indices*.

Activity 1.1

Shorthand form

Write down the shorthand form for:

1. $6 \times a$

2. $2 \times a \times 5 \times b$

3. $b \times 3 \times 4 \times a \times c$

4. $a \times 2 \times b \times a \times c$

1.5 Algebraic processes

Substitution (evaluation)

In the process of substitution we replace a given letter by a numeral. This can best be seen and understood with the help of some activities.

Activity 1.2

Substituting values for variables in algebraic expressions

If $a=2$ and $b=5$, find the values of:

1. $6a$

2. $2a+b$

3. $3ab$

4. a^2+b^2

Activity 1.3

Evaluation by substitution of a well-known formula

Use the formula for the volume of a cuboid: $v=lbh$ with length l cm, breadth b cm and height h cm. Find the volume of a particular box with length 4cm, breadth 6cm and height 2cm by substituting the values 4, 6 and 2 for l, b and h respectively to obtain the volume.

Even if one is using a calculator for basic arithmetic, it is good policy to simplify any algebraic expression as much as possible before substituting values for the numbers.

Simplification – gathering like terms

In the expression $a+3a+2b-2a+6b$, the 'like terms' are of two kinds: the a, the $3a$ and the $-2a$ can be gathered together on the one hand, and the $2b$ and $6b$ gathered together on the other, so that we get:

$a+3a+2b-2a+6b=(a+3a-2a)+(2b+6b)=2a+8b$.

In the expression $2+x+x^2+5x-9+3x^2$ on the other hand, the 2 and -9 are like terms, as are separately the x and the $5x$, and separate from the x^2 and the $3x^2$. Reforming the expression we get $4x^2+6x-7$, which is the simplified expression.

Now try some for yourself.

Activity 1.4

Simplifying algebraic expressions by collecting like terms

Simplify the following expressions:

1. $a+3a+2b-2a+6b$

2. $1+7a-3b-7b+4a-8$

3. $2+x+x^2+5x-9+3x^2$

4. $ab+8a-7b+5ab-2a$

5. $ac+7c+5ca-6c+5a$

6. $4ab+bc+5bc+3a+8-2cb$

7. $3c^2+2cb-c^2+4bc+9c$

In the simplifications that you have just been doing you will have noticed the use of brackets. Let's now give these a bit more attention.

1.6 Brackets

If we want to multiply a whole expression by a number, or another expression, or to subtract it from another expression, or to raise it to some power, it is convenient to enclose each expression in a pair of brackets.

Printers distinguish between parentheses ( ), brackets [ ] and braces { }.

Mathematicians refer to all three kinds as **brackets** and often use the different kinds to distinguish the different levels if they need brackets within brackets.

Computer algebra systems (such as DERIVE) and spreadsheets use ( ) while keeping [ ] for a special purpose. { } is used in mathematics and in computer algebra systems to denote a *set* (see chapter 5).

The removal of brackets while retaining the correct interpretation of the symbols is called **expanding**, but will often be referred to as **simplifying**. Usually, less arithmetic is involved in substitution if the expression inside the brackets is evaluated first.

As in arithmetic 5(2+98) means 5x100, so in algebra $3(a+6a+5b)$ simplifies to $3(7a+5b)$ and means $21a+15b$. Simplify as much as you can inside the bracket before multiplying each term, that is, $7a$ and $5b$, by 3.

Two brackets can be multiplied together thus:

$(a+b)(c+d)=ac+ad+bc+bd.$

In $(x+3)(2x-5)$ we get $2x^2+6x-5x-15$ which contains three types of terms: x squared, x and numeral.

We gather like terms together to get $2x^2+x-15$.

Notice that just as 6-5=1 so $6x-5x=1x$.

Brackets are also useful for holding together a complicated denominator.

Be aware that a fraction such as $\frac{3}{a+b}$ contains an implied bracket in that this expression could also be written as $3/(a+b)$ or $3÷(a+b)$.

You will recall the comment earlier that the order of multiplication does not matter. Thus, for example, $ab=ba$. The same rule still holds when the expressions have brackets. For example, $(a+b)c=c(a+b)=ac+bc$.

Activity 1.5

Simplifying algebraic expressions by removing brackets ('expanding brackets')

Expand (or simplify) the following expressions:

1. $3(a+6)$

2. $4(a+b)$

3. $2(3a-7b)$

4. $a(a+5b+8)$

5. $(3a)^2$

6. $(2a+9b)/3$

7. $(3a+9b)-(2a+5b)$

8. $(a+b)(a-b)$

9. $(x-2)(x+3)$

1.7 Highest common factor (HCF)

In the section, *Brackets,* we set out to 'simplify' algebraic expressions by removing brackets. Sometimes, we 'simplify' by putting brackets in! We may also want to do this in order to make an expression easier to deal with in certain situations. (An example of this is 'completing the square', which you will meet in chapter 3 on *quadratics*.)

Recall that a whole number which divides a given number exactly is called a **factor** of that number. For example, 6 is a factor of 24 but 7 is not.

A **common factor** of two given numbers is a whole number which divides each of those numbers exactly. So, for example:

- 2 is a common factor of 16 and 24
- 4 is a common factor of 16 and 24
- 8 is a common factor of 16 and 24.

The largest such common factor is 8 so this is called the **highest common factor** of 16 and 24. Highest common factor is often abbreviated to HCF.

Activity 1.6

Finding the highest common factor of numerals

Find the HCFs of the following:

1. 18, 24

2. 18, 24, 45

Activity 1.7

Finding the highest common factor of algebraic expressions

Find the HCFs of the following:

1. $2ax, 3bx$

2. ab^2, a^2b

3. $6xy^2, 12x^2y, 18xy$

1.8 Factorising expressions

Factorising can be thought of as the opposite operation of expanding. Here, we put brackets in rather than take them out. One purpose is to make the expression easier to evaluate.

Example: Consider the expression $6xy+18^2y$.

We see that the HCF of the two terms $6xy$ and $18y^2$ is $6y$ which means that we can write:

$$6xy+18y^2=6y.x+6y.3y=6y(x+3y)$$

which is the factorisation.

Factorising algebraic expressions

Factorise the following expressions:

1. $3a+6b$

2. $ax+ay$

3. $pqr+pqs$

Factorising more complicated expressions

Suppose we want to factorise the expression $ac+bc+ad+bd$. It is not at all clear at first what the factors might be, or even if there are any.

For a start, there is no one number which is common to all four terms of the expression.

However, we do see that 'c' is a factor of the first two terms and 'd' is a factor of the last two terms. Therefore, let's try this factorisation for a start:

$$ac+bc+ad+bd=(a+b)c+(a+b)d.$$

Now we effectively have two terms, namely, $(a+b)c$ and $(a+b)d$. We note that $(a+b)$ is a factor of each term, so is a factor of the whole expression. We then get:

$$ac+bc+ad+bd \qquad =(a+b)c+(a+b)d$$
$$=(a+b)(c+d)$$

which is the factorisation.

Factorising more complicated algebraic expressions

Factorise the following expressions:

1. $ax+bx+ay+by$

2. $ab+b^2+ac+bc$

3. $pr-ps+qr-qs$

1.9 **Working with fractions**

Numerical

Example of multiplying fractions:

$$\frac{4}{15}\times\frac{9}{8}=\frac{4\times1}{3\times5}\times\frac{3\times3}{4\times2}=\frac{1}{5}\times\frac{3}{2}=\frac{1\times3}{5\times2}=\frac{3}{10}$$

Example of dividing fractions:

$$\frac{2x}{3y}\div\frac{6y}{15x}=\frac{2x}{3y}\times\frac{15x}{6y}=\frac{2\times x}{3\times y}\times\frac{3\times5x}{2\times3y}=\frac{x}{y}\times\frac{5x}{3y}=\frac{5x^2}{3y^2}$$

(Dividing by a number is the same as multiplying by its reciprocal.)

Example 1 of adding fractions:

$$\frac{a}{b}+\frac{c}{d}=\frac{a\times d}{b\times c}+\frac{c\times b}{d\times b}=\frac{ad}{bd}+\frac{bc}{bd}=\frac{ad+bc}{bd}$$

Example 2 of adding fractions:

$$\frac{4}{15}+\frac{5}{6}=\frac{8}{30}+\frac{25}{30}=\frac{33}{30}$$

The smallest number that 15 and 6 both divide exactly into is 30. (This is called the **lowest common multiple** – LCM for short.)

Algebraic

Example of multiplying fractions:

$$\frac{ax^2}{by}\times\frac{cy^2}{dx}=\frac{ax\times x}{b\times y}\times\frac{cy\times y}{d\times x}=\frac{ax}{b}\times\frac{cy}{d}=\frac{acxy}{bd}$$

Example of dividing fractions:

$$\frac{2x}{3y}\div\frac{6y}{15x}=\frac{2x}{3y}\times\frac{15x}{6y}=\frac{2\times x}{3\times y}\times\frac{3\times 5x}{2\times 3y}=\frac{x}{y}\times\frac{5x}{3y}=\frac{5x^2}{3y^2}$$

Example 1 of adding fractions:

$$\frac{a}{b}+\frac{c}{d}=\frac{a\times d}{b\times d}\times\frac{c\times b}{d\times b}=\frac{ad}{bd}+\frac{bc}{bd}=\frac{ad+bc}{bd}$$

The LCM of b and d is bd.

Example 2 of adding fractions:

$$\frac{p}{q}+\frac{q}{p}=\frac{p\times p}{p\times q}+\frac{q\times q}{q\times p}=\frac{p^2+q^2}{pq}$$

The LCM of p and q is pq.

Activity 1.10

Working with algebraic fractions

Simplify each of the following into a single algebraic fraction:

1. $\dfrac{b}{a}-\dfrac{a}{b}$

2. $\dfrac{a}{b^2}\times\dfrac{b}{a^2}$

3. $\dfrac{a}{b}\div\dfrac{b}{a}$

4. $\dfrac{a}{b}+\dfrac{b}{c}+\dfrac{c}{a}$

1.10 Changing the subject of a formula (also known as transposing)

Example 1

$V=lbh$ is the formula for the volume, V, in cc of a box (cuboid) with length l cm, breadth b cm and height h cm. Suppose that we need to design a box with length 5cm, breadth 3cm and volume 48cc. We now need to make h the subject of the formula instead of V.

First rewrite the formula with the sides changed over. We want h on its own on the left-hand side (LHS).

Complication: h is multiplied by lb – *undo* by dividing by lb. So we get:

$$\frac{V}{lb} = h \qquad \text{giving} \qquad h = \frac{V}{lb}$$

Now making the substitution of the given values for l (= 5), b (= 3) and V (= 48), we get:

$$h = \frac{48}{5 \times 3} = \frac{16}{5} \ (cm)$$

Example 2

$I = \dfrac{PRT}{100}$ is the formula for simple interest.

Suppose we want to know how long it will take to build £500 up to £800 if the interest rate on our money is 10% per annum.

This means we are given values for I, P and R and have to find T.

Make T the subject of the formula.

First rewrite the formula with the sides changed over:

$$\frac{PRT}{100} = I$$

Complication: T is multiplied by PR – *undo* by dividing by PR:

$$\frac{T}{100} = \frac{I}{PR}$$

Complication: T is divided by 100 – *undo* by multiplying by 100:

$$T = \frac{100\,I}{PR}$$

With the given values for I (800-500=300), R (= 10) and P (= 500), we get:

$$T = \frac{100 \times 300}{500 \times 10} = 6 \,(years)$$

Example 3

$S = 2\pi r(r+h)$ is the formula for the surface area of a closed cylinder with radius r cm and height h cm.

To make h the subject, first rewrite the expression as:

$$2\pi r(r + h) = S$$

Complication: Here h is inside a bracket so get the bracket on its own first (by dividing by $2\pi r$).

$$r + h = \frac{S}{2\pi r}$$

Now *h* has *r* added to it, so subtract *r* from both sides:

$$h = \frac{S}{2\pi} - r$$

as required.

Example 4

The formula $a^2 = b^2 + c^2$ is *Pythagoras Theorem* for the length of the hypotenuse of a right-angled triangle.

To make *c* the subject, first rewrite the expression as:

$$b^2 + c^2 = a^2$$

Complication: the LHS doesn't yet have *c* on its own: here *c* is squared, so get *c* squared on its own first by subtracting b^2 from both sides:

$$c^2 = a^2 - b^2$$

Now take the square root of both sides:

$$c = \sqrt{a^2 - b^2}$$

which is the result we want.

Example 5

$$T = 2\pi \sqrt{\frac{l}{g}}$$ is a formula for the time for a pendulum to complete one swing.

To make *g* the subject, before changing sides get rid of the square root by squaring both sides:

$$T^2 = 4\pi^2 \left(\frac{l}{g} \right)$$

Now get rid of the fraction by multiplying both sides by *g* (this actually also gets *g* on the LHS):

$$gT^2 = 4\pi^2 l$$

Now *g* is multiplied by T^2, so we need to **undo** that to get:

$$g = \frac{4\pi^2 l}{T^2}$$

Example 6

$$\frac{1}{u} + \frac{1}{v} = \frac{1}{f}$$ is a formula used in optics.

To make *v* the subject, we isolate $\frac{1}{v}$ (called the *reciprocal* of *v*) first:

$$\frac{1}{v} = \frac{1}{f} - \frac{1}{u}$$

Now we combine the fractions on the right-hand side (RHS) into one fraction:

$$\frac{1}{v} = \frac{u - f}{fu}$$

Now we take reciprocals (turn the fractions upside down) on each side:

$$v = \frac{fu}{u - f}$$

...and we have what we want.

Activity 1.11

Changing the subject of a formula

Transpose each of the following formulas:

1. $x = b/y$, to find y

2. $I = E/R$, to find E

3. $y = mx + c$, to find c

4. $S = ar(r + h)$, to find h

1.11 Indices 1

As in arithmetic, where 3^2 means 3 squared (3 multiplied by 3) so in algebra a^2 means a squared, or a times a. We say we are raising a to the power 2. The 2 in a^2 is called an *index*.

(The origin of the use of the term 3 'squared' lies in finding areas: the area of a square of side 3 units is 3x3 square units, which we write in modern notation as $3^2 = 9$, of course.)

It is important to appreciate that 3x3 and 3^2 and 9 are simply three ways of expressing the number nine.

Similarly, 2 'cubed', for example, arises as the volume of a cube of side 2 units.

As another example: $axaxaxa$ simplifies to a^4 and $2xaxbxcxb$ simplifies to $2ab^2c$.

The rules for indices

How would we multiply two numbers that are expressed in index form? Consider:

$$3^4 \times 3^2 = (3 \times 3 \times 3 \times 3) \times (3 \times 3)$$
$$= 3^6$$

This means that to multiply two numbers written in this way we just add the two indices together to get the product:

$$3^4 \times 3^2 = 3^{4+2} = 3^6$$

We can express this **law** in general terms (that is, using letters instead of numerals) as:

$$a^m \times a^n = a^{m+n}$$

where, for the moment anyway, m and n are positive integers (though a can represent any number at all).

In a similar way, to see how we might divide two numbers written in index form, consider:

$$4^7 \div 4^2 = \frac{4\times4\times4\times4\times4\times4\times4}{4\times4}$$

$$\frac{4\times4\times4\times4\times4}{1\times1}$$

$$\cdot \quad 4^5$$

This means that to divide two numbers written in this way we just subtract one index from the other to get the result:

$$4^7 \div 4^2 = 4^{7-2} = 4^5$$

We can express this **law** in general terms (that is, using letters instead of numerals) as:

$$a^m \div a^n = a^{m-n}$$

where, for the moment anyway, m and n are positive integers (though a can represent any number at all). We shall also require (for the moment) that m is larger than n (written $m>n$): this then means that $m-n$ is a positive number.

(You might like to consider what happens if $m<n$ or, indeed, if $m=n$.)

What now if we want to raise a number in index form to some power? Consider:

$$(2^3)^4 = (2^3)\times(2^3)\times(2^3)\times(2^3)$$

$$= (2\times2\times2)\times(2\times2\times2)\times(2\times2\times2)\times(2\times2\times2)$$

$$= 2^{12}.$$

This shows that to raise a number written in index form to some power we just multiply the two indices together to get the result:

$$(2^3)^4 = 2^{3\times4} = 2^{12}.$$

We can express this **law** in general terms (that is, using letters instead of numerals) as:

$$(a^m)^n = a^{m\times n}$$

Again, m and n are positive integers (for the moment).

Activity 1.12

Evaluating numbers written in index form (where the index is a positive integer)

Evaluate the following:

1. $3^2 \times 3^5$

2. $5^4 \times 5^3$

3. $7^5 \div 7^3$

4. $6^4 / 6$

5. $(2^2)^3$

Activity 1.13

Simplifying algebraic expressions already in index form (where the index is a positive integer)

Simplify the following:

1. $a^2 \times a^3$

2. $a^{17} \times a^5$

3. $b^6 \div b^4$

4. $\dfrac{x^5}{x}$

5. $(a^2)^3$

6. $(a^4)^2$

1.12 Indices 2

So far in the work on indices, we have considered only the cases where the index is a positive integer. We now need to extend this to the following situations:

1. When the index is a negative number.

2. When the index is zero.

3. When the index is a fraction (rational number).

1. When the index is a negative number

What, for example, do we mean when we write something like 2^{-1} or 2^{-3}?

These are further examples of mathematical shorthand. When we write 2^{-1} we mean $\frac{1}{2}$ and when we write 2^{-3} we mean $\frac{1}{2^3}$. That is, the negative in the index means 'take the reciprocal'. So 2^{-1} is the reciprocal of 2 (strictly 2^1, but this is just 2) and 2^{-3} is the reciprocal of 2^3.

2. When the index is zero

Now this is trickier. When we write 2^3, for example, we think of writing down three 2s and multiplying them together. But, when we write 2^0 it is not possible to think of writing down zero 2s and multiplying them together. This is just a nonsense statement. So does 2^0 mean anything at all and if so, what?

Consider the following table showing the structure of a base 10 (decimal) number:

Ten thousands	Thousands	Hundreds	Tens	Units	Tenths	Hundredths
10,000	1,000	100	10	1	$\frac{1}{10}$	$\frac{1}{100}$
10^4	10^3	10^2	10^1		10^{-1}	10^{-2}

The second row of the table gives the position value as a power of ten. You can see the index falling from 4 down to 1, then a jump across the units position, then continuing to fall from -1 to -2. There is a gap in the units position and we might wonder what power of 10 would be appropriate here. If we write down the falling indices as a sequence of numbers, we would get:

$$\ldots,4,3,2,1, \quad ,-1,-2,\ldots$$

and it's easy to see (we hope) that the missing number is 0. So, if we want to continue the pattern of powers of ten, we need to have 10^0 in the units position. We therefore make a **definition:**

$$10^0 = 1$$

More generally, we write:

$$a^0 = 1$$

The rules of indices

These continue to apply when the index is a negative integer, and zero. Thus, we now have the rules applying to all cases where the index is a whole number (**integer**):

$$a^m \times a^n = a^{m+n}$$

$$a^m \div a^n = a^{m-n}$$

$$(a^m)^n = a^{mn}$$

Activity 1.14

Simplifying algebraic expressions already in index form (where the index can be any integer)

Simplify the following:

1. $(a^2)^{-1}$
2. $(a^3)^{-2}$
3. $(x^{-5})^3$
4. $\dfrac{1}{a^4}$
5. x^0

3. When the index is a fraction (rational number)

What, for example, do we mean when we write something like $3^{1/2}$ or $2^{-1/3}$? These are further examples of mathematical shorthand.

If $3^{1/2}$ were to have any meaning, we should want it to fit in with the laws of indices as expressed so far. Suppose, then, that $3^{1/2}$ does have a meaning, and suppose further, that the laws of indices apply, also, when the index is a fraction. Then we should want:

$$3^{1/2} \times 3^{1/2} = 3^{1/2+1/2} = 3^1 = 3$$

But:

$$\sqrt{3} \times \sqrt{3} = 3$$

This means that $3^{1/2}$ will make some sense as a number if we identify it with $\sqrt{3}$. So we **define:**

$$3^{1/2} = \sqrt{3}$$

More generally, we define:

$$a^{1/n} = \sqrt{a}$$

The nth root of a can be written $a^{1/n}$. That is, $a^{1/n} = \sqrt[n]{a}$

The square root of a cubed is written $\sqrt{a^3} = (a^3)^{1/2} = a^{3\times(1/2)} = a^{3/2}$

On the other hand, the cube root of a squared is written $\sqrt[3]{a^2} = (a^2)^{1/3} = a^{2\times(1/3)} = a^{2/3}$

More generally, the nth root of a^m is written $\sqrt[n]{a^m} = a^{m/n}$

We have the following rules of indices which apply, no matter what kind of number the index is:

$$a^m \times a^n = a^{m+n}$$

$$a^m \div a^n = a^{m-n}$$

$$(a^m)^n = a^{mn}$$

Note. In general, a, too, may be any number but we must take care in this case if the index is a fraction. For example, what might we mean by $(-3)^{1/2} = \sqrt{-2}$ (the square root of a negative number)? Does this mean anything? The answer, in fact, is yes – but a bit beyond this course.

Activity 1.15

Evaluating numbers written in index form (where the index can be any number)

Evaluate the following:

1. $16^{1/2}$

2. $8^{1/3}$

3. $27^{2/3}$

4. $9^{-1/2}$

5. 10^0

Activity 1.16

Simplifying algebraic expressions to index form

Write down the shorthand form for:

1. $a \times a \times a \times a$

2. $2 \times a \times b \times b \times c$

3. $p \times p \times 3 \times q \times q \times q \times r$

Activity 1.17

Highest common factors with indices

1. Find the highest common factor of $2^3 \times 3^2$, $2^2 \times 3^4$ expressing your answer in index form.

2. Find the HCFs of the following: $6xy^2$, $12x^2y$, $18x^3y^3$.

1.13 Summary

In this chapter we saw how to use the language of algebra to say things about numbers in an organised way. We saw how to do calculations, change the subject of a formula, and solve problems using algebra. These ideas will be used throughout the rest of the book.

1.14 Review questions

 Question 1.1: Evaluation by substitution

If $a=3$ and $b=4$, find the values of:

1. $4a$

2. a^2

3. ab

4. $5a^2b$

5. $2a+3b$

6. a^2; a^3; a^4

7. $\sqrt{a^2 + b^2}$

8. $\sqrt{ab}$

Question 1.2: Simplifying by gathering like terms

1. Collect like terms in the following:

   $3a-4b+2c-6b+2a-5c$

2. Collect like terms in the following:

   $-6pq+14pq-5pq$

3. Add up the following:

   $3x-2y+7z$

   $-3x-4y+3z$

   $5x+3y-11z$

4. Add up the following:

$2x^2-3x+5$

$-2x^2-4x-2$

$-7x^2+2x-6$

Question 1.3: Simplifying by expanding brackets

Expand (or simplify) the following expressions:

1. i. $2a(3a+4b+9)$

   ii. $(x-7y)-(2x-3y)$

   iii. $(18a)/(3b)$

   iv. $(x+y)\left(\dfrac{1}{x}+\dfrac{1}{y}\right)$

   v. $(a+b)^2$

   vi. $(a+b)^3$

2. Remove the brackets and simplify:

   i. $5(x-3)-2(2x+3)$

   ii. $x(x+3)-3x$

   iii. $x(x-10)+4(x-10)$

   iv. $x(x+5)-5(x+5)$

   v. $a(x+2)$

   vi. $(x+3)(x+2)$

   vii. $(x-1)b$

   viii. $(x-1)(x+4)$

Question 1.4:

1. Show that $(x+y)^2-(y+z)^2+(x-z)^2 = 2(x+y)(x-z)$.
2. If $a=\dfrac{1}{x}$ and $c=\dfrac{1}{1-y}$, show that $y=1-a$.

Question 1.5: Finding the highest common factor (HCF) of numbers

Find the HCFs of the following:

1. 120, 80, 90
2. 30, 42, 12
3. $2^4\times5,2^3\times5^4,2^5\times5^2\times9$, expressing your answer in index form

Question 1.6: Finding the highest common factor (HCF) of algebraic expressions

Find the HCFs of the following:

1. abc^2, a^2bc, ab^2c
2. $36y^2$, $-24y^3$, $18y^5$
3. $z^4(x^2+1)^3$, $z^5(x^2+1)^2$

Question 1.7: Factorising algebraic expressions

Factorise the following expressions:

1. $9a+6b$
2. $pqr+pqs$
3. $2x^2y+2xy^2$
4. $3x+6y-3z$
5. $ac+bc+ad+bd$
6. $a-a^2b-b+ab^2$
7. x^2+5x+6
8. $2a^2-2ab-ac+bc$
9. $4(x-1)^2-8(x-1)-5$
10. $x^2+y^2-a^2-2xy$
11. $(5x+4y)^2-(3x+y)^2$
12. x^8-1.

Question 1.8: Working with algebraic fractions

Simplify the following:

1. $\dfrac{a^2b}{ab}$

2. $\dfrac{xyz}{y^2}$

3. $\dfrac{18p}{q} \times \dfrac{p}{8q}$

4. $\dfrac{2m}{n} \div \dfrac{n}{2m}$

5. $\dfrac{x}{y} \times \dfrac{z}{w} \div \dfrac{xz}{yw}$

6. $\dfrac{a^2+ab}{b} \times \dfrac{c}{ac+ab}$

7. $\dfrac{1}{2a} + \dfrac{2}{b}$

8. $\dfrac{1}{x+y} + \dfrac{1}{x-y}$

9. $\dfrac{x+y}{x-y} + \dfrac{x-y}{x+y}$

Question 1.9: In an electrical circuit a resistance of R ohms is placed in series with n batteries each of internal resistance r ohms. Each battery produces an emf of E volts. The formula for the current I (amps) flowing in the circuit of such an arrangement is given by the formula:

$$I = \frac{nE}{R + nr}$$

If R = 25ohms, n = 5, E = 2 volts, r = 1.3ohms, find the value of I.

Question 1.10: Changing the subject of a formula

1. The formula $PV=nRT$ is used in connection with the study of gases. Make P, V, R and T, in turn, the subject of the formula.

2. The formula $v = \sqrt{2gh}$ is used in physics. Make g the subject of the formula.

3. The formula $v=u+at$ is used in physics. Make a the subject.

4. The formula $v=u^2+2as$ is used in physics. Make a the subject of the formula.

5. Rearrange the expression to make its subject the letter shown in the square bracket:

   $A = 2\pi rh$ [h].

6. Rearrange the expression to make its subject the letter shown in the square bracket:

   $V = \frac{4}{3}\pi r^3$ [r].

7. Rearrange the expression to make its subject the letter shown in the square bracket:

   $B = 2b + \sqrt{b^2 + 6bu}$ [u].

Question 1.11

If $V = 2\pi r^2 h$ and $A = 2\pi rh$, find V in terms of A and r.

Question 1.12: Simplifying powers of numbers 1

1. Simplify:

   i. $a^2 \times a^3$ ii. $a^5 \times a^4$ iii. $a^7 \times a^2$

   Can you find a general result? What is $a^m \times a^n$ for m, n positive integers?

2. Simplify:

   i. $a^5 \div a^3$ ii. $a^8 \div a^5$ iii. $a^2 \div a^3$

   Can you find a general result? What is $a^m \div a^n$ for m, n positive integers?

3. Thus write down the result of $a^m \times a^n$ for m, n any integers.

   Similarly for $a^m \div a^n$. What about $a^m \times a^{-n}$?

Question 1.13: Simplifying powers of numbers 2

1. What is another way of writing the following?

   i. a^0 ii. $a^{1/2}$ iii. $a^{2/3}$ iv. $a^{1/n}$

2. Simplify:

   i. $(a^m)^p$ ii. $(a^m)^{1/q}$ iii. $(a^m)^{p/q}$

   where in all cases m, n are positive integers.

3. Simplify each of the following by expressing as a single algebraic fraction:

   i. $\sqrt{a} + \dfrac{b}{\sqrt{a}}$ ii. $\sqrt{a+b} - \dfrac{a}{\sqrt{a+b}}$

 Question 1.14: When $x = -\frac{1}{2}$, find the value of:

1. x^2

2. $\dfrac{1}{x}$

3. $\dfrac{1}{x^2}$

4. $\left(\dfrac{x}{2}\right)^{-2}$

5. $(-x)^{-1}$

6. $x^{2/3}(27x)^{-1/3}$

 Question 1.15: A square rug of side x metres is placed in the middle of a square room of side y metres.

1. Find the area of the floor not covered by the rug.

2. If this area is equal to half the area of the rug, show that $y = \sqrt{\dfrac{3}{2}x}$.

   (Don't forget that $\sqrt{a \times b} = \sqrt{a} \times \sqrt{b}$.)

 **Question 1.16:** If an amount of money P is invested in an account that earns $i\%$ per year compound interest, the amount A in the account after n years is given by the formula:

$$A = P\left(1 + \frac{i}{100}\right)^n$$

If £1,000 is invested in this account, what is the value of the investment after 5 years if the interest rate is 4% per year?

1.15 Answers to review questions

Question 1.1: Evaluation by substitution

1. $4 \times 3 = 12$
2. $3^2 = 3 \times 3 = 9$
3. $3 \times 4 = 12$
4. $5 \times 3^2 \times 4 = 5 \times 9 \times 4 = 20 \times 9 = 180$
5. $2 \times 3 + 3 \times 4 = 6 + 12 = 18$
   Remember, if there are no brackets present, do multiplication before addition.
6. $3^2 = 9$; $3^3 = 27$; $3^4 = 18$
7. $\sqrt{3^2 + 4^2} = \sqrt{25} = 5$

   (In geometry: the 3, 4, 5 right-angled triangle – an application of Pythagoras' Theorem)
8. $\sqrt{3 \times 4} = \sqrt{3} \times \sqrt{4} = \sqrt{3} \times 2 = 2\sqrt{3}$

Question 1.2: Simplifying by gathering like terms

1. $5a\text{-}10b\text{-}3c$
2. $3pq$
3. $5x\text{-}3y\text{-}z$
4. $\text{-}7x^2\text{-}5x\text{-}3$

Question 1.3: Simplifying by expanding brackets

1. i. $6a^2+8ab+18a$ ii. $\text{-}x\text{-}4y$
   iii. $6/b$ iv. $\dfrac{x}{y}+\dfrac{y}{x}+2$
   v. $a^2+2ab+b^2$ vi. $a^3+3a^2b+3ab^2+b^3$

2. i. $x\text{-}21$ ii. x^2
   iii. $x^2\text{-}6x\text{-}40$ iv. $x^2\text{-}25$
   v. $ax+2a$ vi. x^2+5x+6
   vii. $bx\text{-}b$ viii. $x^2+3x\text{-}4$

Question 1.4

1. Just expand the brackets, collect like terms, then factorise the resultant expression.
2. First note that $\dfrac{1}{x} = 1\text{-}y$, so $1\text{-}\dfrac{1}{x} = y$. The rest follows readily enough.

Question 1.5: Finding the highest common factor (HCF) of numerals

1. 10

2. 6

3. $2^3 \times 5$

Question 1.6: Finding the highest common factor (HCF) of algebraic expressions

1. abc

2. $6y^2$

3. $z^4(x^2+1)^2$

Question 1.7: Factorising algebraic expressions

1. $3(3a+2b)$

2. $pq(r+s)$

3. $2xy(x+y)$

4. $3(x+2y-z)$

5. $ac+bc+ad+bd = (a+b)c+(a+b)d = (a+b)(c+d)$

6. $a-a^2b-b+ab^2 = a(1-ab)-b(1-ab) = (a-b)(1-ab)$

7. $(x+2)(x+3)$

We know that $(x+a)(x+b) = x^2+(a+b)x+ab$ so we are looking for two numbers that multiply to give 6 and that add to give 5. The required numbers are 2 and 3.

8. $(2a-c)(a-b)$

9. $(2x-7)(2x-1)$

10. $(x-y+a)(x-y-a)$

11. $(8x+5y)(2x+3y)$

12. $x^8-1=(x^4-1)(x^4+1)=(x^2-1)(x^2+1)(x^4+1)=(x-1)(x+1)(x^2+1)(x^4+1)$

Question 1.8: Working with algebraic fractions

1. $\dfrac{a^2b}{ab} = \dfrac{a \times ab}{ab} = a$

2. $\dfrac{xyz}{y^2} = \dfrac{y \times xz}{y \times y} = \dfrac{xz}{y}$

3. $\dfrac{18p}{q} \times \dfrac{p}{8q} = \dfrac{2 \times 9p}{q} \times \dfrac{p}{2 \times 4q} = \dfrac{9p \times p}{q \times 4q} = \dfrac{9p^2}{4q^2}$

4. $\dfrac{2m}{n} \div \dfrac{n}{2m} = \dfrac{2m}{n} \times \dfrac{2m}{n} = \dfrac{4m^2}{n^2}$

5. $\dfrac{x}{y} \times \dfrac{z}{w} \div \dfrac{xz}{yw} = \dfrac{xz}{yw} \times \dfrac{yw}{xz} = 1$

6. $\dfrac{a^2+ab}{b} \times \dfrac{c}{ac+bc} = \dfrac{a(a+b)}{b} \times \dfrac{c}{c(a+b)} = \dfrac{a}{b}$

7. $\dfrac{1}{2a} + \dfrac{2}{b} = \dfrac{b}{2ab} + \dfrac{4a}{2ab} = \dfrac{b+4a}{2ab}$

8. $\dfrac{1}{x+y} + \dfrac{1}{x-y} = \dfrac{x-y}{(x+y)(x-y)} + \dfrac{x+y}{(x+y)(x-y)} = \dfrac{(x-y)+(x+y)}{x^2-y^2} = \dfrac{2x}{x^2-y^2}$

9. $\dfrac{x+y}{x-y} + \dfrac{x-y}{x+y} = \dfrac{(x+y)(x+y)}{(x-y)(x+y)} + \dfrac{(x-y)(x-y)}{(x-y)(x+y)} = \dfrac{(x+y)^2+(x-y)^2}{x^2-y^2} = \dfrac{2(x^2+y^2)}{x^2-y^2}$

Question 1.9

$I = \dfrac{10}{31.5} = 0.3$ amps (to one decimal place).

Question 1.10: Changing the subject of a formula

1 $P = \dfrac{nRT}{V}$ $\quad V = \dfrac{nRT}{P}$ $\quad R = \dfrac{PV}{nT}$ $\quad T = \dfrac{PV}{nR}$

2. $g = \dfrac{V^2}{2h}$ $\quad$ First square each side; then divide each side by 2*h*.

3. $a = \dfrac{v-u}{t}$ $\quad$ First subtract u from each side; then divide each side by *t*.

4. $a = \dfrac{v^2-u^2}{2s}$ First subtract *u*^{2} from each side; then divide each side by 2*s*.

5. $h = \dfrac{A}{2\pi r}$

6. $r = \left(\dfrac{3V}{4\pi}\right)^{1/3}$

7. $B - 2b = \sqrt{b^2 + 6bu}$ $\quad$ so, squaring, (B-2b)²=b²+6*bu*, which, after further rearrangement, gives:

$u = \dfrac{(B-2b)^2 - b^2}{6b} = \dfrac{(B-b)\times(B-3b)}{6b}$

Question 1.11

Eliminate *h* between the two formulas: $h = \dfrac{V}{\pi r^2} = \dfrac{A}{2\pi r}$, so $V = \dfrac{\pi r^2 \times A}{2\pi r} = \dfrac{\pi r \times r \times A}{\pi r \times 2} = \dfrac{rA}{2}$

Question 1.12: Simplifying powers of numbers 1

1. i $\quad a^5$

   ii $\quad a^9$

   iii $\quad a^9$

These results suggest the general result: $a^m \times a^n = a^{m+n}$, where *m*, *n* are positive integers.

2. i $\quad a^2$

   ii $\quad a^3$

   iii $\quad a^1$

These results suggest the general result: $a^m \div a^n = a^{m-n}$, where *m*, *n* are positive integers.

3. For *m*, *n* any integers (not just positive integers) we have the general result for multiplication of two powers:

$$a^m \times a^n = a^{m+n}$$

And for division of two powers:

$$a^m \div a^n = a^{m-n}$$

Question 1.13: Simplifying powers of numbers 2

1. i.   1

   ii. $\sqrt{a}$

   iii. $\sqrt[3]{a^2} = \left(\sqrt[3]{a}\right)^2$

   iv. $\sqrt[n]{a}$

2. i. a^{mp}

   ii. $a^{m/q}$

   iii. $a^{mp/q}$

3. i. $\dfrac{a+b}{\sqrt{a}}$

   ii. $\dfrac{b}{\sqrt{a+b}}$

Question 1.14

1. $\dfrac{1}{4}$

2. -2

3. 4

4. 16

5. 2

6. $-\dfrac{3}{2}$

Question 1.15

1. No need for anything fancy here. Just note that the area of the room is y^2 and the area of the rug is x^2, so the area uncovered must be y^2-x^2.

2. We require $y^2-x^2 = \frac{1}{2}x^2$ which gives $y^2 = \frac{3}{2}x^2$ and the result follows. (Note that y must be positive.)

Question 1.16

$A = 1{,}000(1.04)^5 = £1{,}216.65$ (to 2 decimal places: for currency).

[So the total interest earned = £216.65.]

1.16 Feedback on activities

Activity 1.1

1. $6 \times a = 6a$
2. $2 \times a \times 5 \times b = 10ab$
3. $b \times 3 \times 4 \times a \times c = 12abc$. We commonly write the letters in the order they appear in the (Latin) alphabet.
4. $a \times 2 \times b \times a \times c = 2a^2bc$

Activity 1.2: Substituting values for variables in algebraic expressions

1. $6a = 6 \times 2 = 12$
2. $2a+b = 2 \times 2+5 = 4+5 = 9$
3. $3ab = 3 \times 2 \times 5 = 30$
4. $a^2+b^2 = 2^2+5^2 = 2 \times 2+5 \times 5 = 4+25 = 29$

Note the order of the operations in 2.: $\times$ is performed before $+$.

Activity 1.3: Evaluation by substitution of a well-known formula

$V = lbh = l \times b \times h = 4\text{cm} \times 6\text{cm} \times 2\text{cm} = 48\text{cm}^3$, which is also written 48cc.

Activity 1.4: Simplifying algebraic expressions by removing brackets

1. $a+3a+2b-2a+6b = (a+3a-2a)+(2b+6b) = 2a+8b$
2. $1+7a-3b-7b+4a-8 = 11a-10b-7$
3. $2+x+x^2+5x-9+3x^2 = -7+6x+4x^2$
4. $ab+8a-7b+5ab-2a = 6a-7b+6ab$
5. $ac+7c+5ca-6c+5a = 5a+c+6ac$. Note that $ca = ac$.
6. $4ab+6c+5bc+3a+8-2cb = 3a+6c+8+4ab+5bc$. Note that $cb = bc$.
7. $3c^2+2cb-c^2+4bc+9c = 2c^2+6bc+9c$. Note that $cb = bc$.

Activity 1.5: Simplifying algebraic expressions by collecting like terms

1. $3(a+6) = 3 \times a+3 \times 6 = 3a+18$
2. $4(a+b) = 4 \times a+4 \times b = 4a+4b$
3. $2(3a-7b) = 2 \times (3a)+2 \times (-7b) = 2 \times 3 \times a+2 \times (-7) \times b = 6a-14b$
   Recall that $(+) \times (-) = (-)$ so that $2 \times (-7) = -2 \times 7$.
4. $a(a+5b+8) = a \times a+a \times 5b+a \times 8 = a^2+5ab+8a$
   Remember that in algebraic expressions we usually like to write the numeral part out in front.
5. $(3a)^2 = (3a) \times (3a) = (3 \times a) \times (3 \times a) = (3 \times 3) \times (a \times a) = 9a^2$
6. $(2a+9b)/3 = 2a/3+9b/3 = \frac{2}{3}a+3b$
7. $(3a+9b)-(2a+5b) = 3a+9b-2a-5b = (3a-2a)+(9b-5b) = a+4b$
8. $(a+b)(a-b) = a \times a+a \times (-b)+b \times a+b \times (-b) = a^2-ab+ab-b^2 = a^2-b^2$

We note:

a) $(+)\times(-) = (-)$ so that $a\times(-b) = -a\times b = -ab$;

b) $ba = ab$.

9. $(x-2)(x+3) = xxx+xx3-2xx-2\times3 = x^2+3x-2x-6 = x^2+x-6$

Activity 1.6: Finding the highest common factor of numerals

1. $18 = 6\times3$ and $24 = 6\times4$, so the largest number that divides exactly into both 18 and 24 is 6. Thus, the HCF of 18 and 24 is 6.

2. $18 = 3\times6$, $24 = 3\times8$ and $45 = 3\times15$, so the largest integer that divides exactly into all three numbers is 3.

Activity 1.7: Finding the highest common factor of algebraic expressions

1. $2ax = x\times2a$, $3bx = x\times3b$, so the HCF is x.

2. $ab^2 = ab.b$, $a^2b = ab.a$, so the HCF is ab.

3. $6xy^2 = 6xy.y$, $12x^2y = 6xy.2x$, $18xy = 6xy.3$. The HCF is $6xy$.

Activity 1.8: Factorising algebraic expressions

1. $3a+6b$: HCF is 3. So, $3a+6b = 3.a+3.2b = 3(a+2b)$.

2. $ax+ay$: HCF is a. So, $ax+ay = a.x+a.y = a(x+y)$.

3. $pqr+pqs$: HCF is pq. So, $pqr+pqs = pq.r+pq.s = pq(r+s)$.

Activity 1.9: Factorising more complicated algebraic expressions

1. $ax+bx+ay+by = (x+b)x+(a+b)y = (a+b)(x+y)$.

2. $ab+b^2+ac+bc = (a+b)b+(a+b)c = (a+b)(b+c)$.

3. $pr-ps+qr-qs = p(r-s)+q(r-s) = (p+q)(r-s)$.

Activity 1.10: Working with algebraic fractions

1. $\dfrac{b}{a}-\dfrac{a}{b} = \dfrac{b^2}{ab}-\dfrac{a^2}{ab} = \dfrac{b^2-a^2}{ab}$

2. $\dfrac{a}{b^2}\times\dfrac{b}{a^2} = \dfrac{a\times1}{b\times b}\times\dfrac{b\times1}{a\times a} = \dfrac{1}{b}\times\dfrac{1}{a} = \dfrac{1}{ab}$

3. $\dfrac{a}{b}\div\dfrac{b}{a} = \dfrac{a}{b}\times\dfrac{a}{b} = \dfrac{a^2}{b^2}$

4. $\dfrac{a}{b}+\dfrac{b}{c}+\dfrac{c}{a} = \dfrac{a\times ac}{abc}+\dfrac{b\times ab}{abc}+\dfrac{c\times bc}{abc} = \dfrac{ca^2+ab^2+bc^2}{abc}$

Activity 1.11: Changing the subject of a formula

1. $x = b/y = \dfrac{b}{y}$: Multiply each side by y: we then get $xy = b$.

   Now divide each side by x to give $y = \dfrac{b}{x}$, which is the new formula that we want.

2. $I = E/R = \dfrac{E}{R}$: Multiply each side by R: we then get $IR = E$ so $E = IR$, which is what we want.

3. $y = mx+x$ so $y-x = mx$ so $\dfrac{y-x}{m} = x$ so $x = \dfrac{y-x}{m}$, which is what we want.

4. $S = ar(r+h)$ so $S = ar^2+arh$ so $S-ar^2 = arh$ so $\dfrac{S-ar^2}{ar} = h$ so $h = \dfrac{S-ar^2}{ar}$

   Alternatively: $S = ar(r+h)$ so $\dfrac{S}{ar} = r+h$ so $\dfrac{S}{ar}-r = h$ so $h = \dfrac{S}{ar}-r$, which is the same thing.

Activity 1.12: Evaluating numbers written in index form

1. $3^2 \times 3^5 = 3^{2+5} = 3^7 = 2{,}187$
2. $5^4 \times 5^3 = 5^{4+3} = 5^7 = 78{,}125$
3. $7^5 \div 7^3 = 7^{5-3} = 7^2 = 49$
4. $6^4/6 = 6^4/6^1 = 6^{4-1} = 6^3 = 216$
5. $(2^2)^3 = 4^3 = 64$

Activity 1.13: Simplifying algebraic expressions

1. $a^2 \times a^3 = a^{2+3} = a^5$
2. $a^{17} \times a^5 = a^{17+5} = a^{22}$
3. $b^6 \div b^4 = b^{6-4} = b^2$
4. $x^5/x = x^{5-1} = x^4$
5. $(a^2)^3 = a^{2 \times 3} = a^6$
6. $(a^4)^2 = a^{4 \times 2} = a^8$

Activity 1.14: Simplifying algebraic expressions

1. $(a^2)^{-1} = \dfrac{1}{a^2} = a^{-2}$. Alternatively, $(a^2)^{-1} = a^{2 \times (-1)} = a^{-2}$.
2. $(a^3)^{-2} = a^{3 \times (-2)} = a^{-6}$
3. $(x^{-5})^3 = x^{(-5) \times 3} = x^{-15}$
4. $\dfrac{1}{a^4} = a^{-4}$
5. $x^0 = 1$

Activity 1.15: Evaluating numbers written in index form

1. $16^{1/2} = 4$. Ask yourself the question: what squared is equal to 16? The answer is 4.
2. $8^{1/3} = 2$. Ask yourself the question: what cubed is equal to 8? The answer is 2.
3. $27^{2/3} = 27^{2 \times (1/3)} = 27^{(1/3) \times 2} = 27^{(1/3)2} = 3^2 = 9$
4. $9^{-1/2} = \dfrac{1}{9^{1/2}} = \dfrac{1}{3}$. Recall that a negative index means the reciprical ('one over ...').
5. $10^0 = 1$. Recall that any number to the power zero has the value 1. This is a **definition**.
   Thought: What is the value of 0^0?

Activity 1.16: Simplifying algebraic expressions in index form

1. $a \times a \times a \times a = a^4$
2. $2 \times a \times b \times b \times c = 2ab^2c$. Note how compact this simplified algebraic form is.
3. $p \times p \times 3 \times q \times q \times q \times r = 3p^2q^3r$

Activity 1.17: Highest common factors with indices

1. $2^3 \times 3^2 = (2^2 \times 3^2) \times 2$ and $2^2 \times 3^4 = (2^2 \times 3^2) \times 3^2$ so the HCF is $2^2 \times 3^2$.
2. $6xy^2 = 6xy \cdot y$, $12x^2y = 6xy \cdot 2x$, $18x^3y^3 = 6xy \cdot 3x^2y^2$, so the HCF is $6xy$.

CHAPTER 2

Linear equations

OVERVIEW

Algebra is one of the oldest subjects in mathematics and is used everyday in countless situations, most of the time without you even knowing. The main idea behind algebra is to use letters to represent familiar objects, like numbers. Numbers satisfy certain conditions. For example it doesn't matter which order you multiply two numbers together, as you get the same answer both ways, so algebra uses letters to represent numbers because then you don't need to worry too much about what the number is, just that it is a number.

In the last chapter you found out how to manipulate algebraic expressions and how different operations, like x, + and - have different priorities. In this chapter, you will put the knowledge you learned in the first chapter to use by solving equations.

Learning outcomes On completion of this chapter, you should be able to:

- Solve linear equations

- Relate two quantities using a formula

- Change the subject of a formula

- Solve written problems using algebra.

2.1 Introduction

You will have learned the rules for manipulating numbers in the first chapter. For example, you will understand what is meant when 2x4+3 or 2x(4+3) is written.

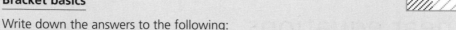

Activity 2.1

Bracket basics

Write down the answers to the following:

- 2x4+3

- 2x(4+3)

Example 2.1

Think about the following sentence: 'When I retire in 40 years, I will be three times as old as I am now.' How old am I?

In this example we need to work out my age given various different pieces of information.

Algebra lets us simplify the information that we are given so that we can solve this problem. Before the end of this chapter you will be able to solve this and find out how old I am. In fact by the end of this chapter you will be able to tell me additional things about my age. For example, you will be able to say how old I will be when I retire.

2.2 Basic manipulation

Just as with numbers you saw in the first chapter, the way in which you write an algebraic expression is vitally important.

Example 2.2

The following two expressions are not the same:

- $2a+3$
- $2(a+3)$

Activity 2.2

Bracket basics - variables

You should spend a few minutes thinking about what the difference is in the two expressions above. Try substituting numbers for the letter a to see what you get. Is there a number you can think of for a that makes the two expressions the same?

Activity 2.3

More on brackets

Which of the following expressions are the same?

1. $2x+x-2$

2. $2(x-1)+x$

3. $x-2$

2.3 Equations

In this section you will learn to understand equations. In particular, you will find out how to write down the sentence in example 2.1, using algebra.

The equals sign

The first and most important thing is the equals sign. The equals sign, written =, means that two quantities are the same. An equation has the format:

> *expression = expression*

Since expressions represent numbers, this means that the two expressions are the same number.

Example 2.3

The following are examples of equations involving numbers only:

- 2x3+4 = 10
- 3(8-3)+2 = 17
- (9-3)/3 = 5-3.

Activity 2.4

Equations

Go through each of the equations in example 2.3 and check that the number on the left is the same as the number on the right.

Example 2.4

The following are examples of equations involving letters and numbers:

- $2x = 12$.

This equation means that if you multiply the number x by 2, then it is 12.

- $2a = 3b$.

This equation means that if you multiply a by 2, and b by 3, then the answers will be the same.

You can see in the example above that equations are a way of representing information about numbers we do not yet know. Later on in this chapter you will find out how you can solve equations to find out what the letters represent. So algebra is a way of writing information down, involving numbers you do not yet know, and then using it to find out the unknown number.

Activity 2.5

Equations as sentences

Write down in sentences, as in example 2.4, what the following equations mean:

- $2a = a+4$
- $3a = 2a-3$

We are now ready to write the sentence in example 2.1 as an equation.

Example 2.5

Think about the sentence we had at the beginning of the introduction: 'When I retire in 40 years, I will be three times as old as I am now.'

First of all we will need to represent the number we are trying to find by a letter. The number we are trying to find is my age; we will call this a. Now we need to convert the sentence into an equation.

The sentence says that my age in 40 years will be the same as my age multiplied by 3. My age in 40 years is given by the expression $a+40$. To understand this a little better, suppose that I am 25 years old. Then in 40 years I will be 25+40 years old, or 65 years old. If I am 30 years old, then in 40 years I will be 30+40 years old, or 70. So if I am a years old now, then in 40 years I will be $a+40$ years old.

If we multiply my age by 3, then we get $3a$. To understand this better, if I am 25 years old now, then three times my age will be 3x25 or 75 years.

The sentence says that these two expressions, $a+40$ and $3a$, are the same. So the sentence can be written as:

$$a+40 = 3a.$$

In the next section we will find out how to solve this equation to find out how old I am.

2.4 Solving equations

So far you have found out how algebra can be used to help solve problems that could otherwise be difficult. To do this, algebra uses letters to represent numbers that we do not know. In the last section you found out that problems can be written as equations in algebra that we can solve. In particular, you saw how the example given at the start of the chapter, to find out my age, could be written as an equation.

In this section you will learn how to solve the equations you saw in the previous section. Once you have learned this, you will be able to solve a great many problems using algebra.

Before you begin, let's refresh our memories about how the various operations of multiplication, addition etc. work and how we represent them using algebra. Remember that arithmetic operations in algebra work just as they do with numbers so, for example, 2+4 is the same as 4+2, but 2-4 is not the same as 4-2.

You will already have learned this in the first chapter, but we will recap briefly how the various operations work so that it is fresh in your mind.

2.5 Multiplying expressions

Example 2.6

Multiply the expression $2a$ by the expression $3b$.

Remember that multiplying algebraic expressions involving letters is the same as multiplying numbers (since the letters are only numbers that we do not yet know). So, we can write:

$$2ax3b \quad = 2 \times 3 \times a \times b$$

$$= 6ab$$

Notice that we do not need the x symbol. If two letters are written next to each other in algebra (like ab), then this means we are multiplying them. So ab means $a \times b$. The order in which we multiply does not matter, $ab = ba$. This is just like with numbers, $2 \times 4 = 4 \times 2$.

So, we could have written the answer in the example as $6ba$ and it would have been correct.

Activity 2.6

Multiplying variables

Multiply the following pairs of expressions:

1. $4x$ and $3y$

2. $3x$ and $4y$

3. $3y$ and $4x$

4. $3x$ and $4x$

2.6 Adding expressions

Let's try the next example.

Example 2.7

Add the following expressions together:

$$2g+4 \text{ and } 3g\text{-}1$$

To do this we collect *like terms* together. Like terms are terms with the same letters in them. In this example the terms $2g$ and $3g$ are like terms; so are 4 and -1.

If we collect these like terms together we get:

$$(2g+4)+(3g\text{-}1) = (2g+3g)+(4+(\text{-}1))$$

$$= 5g+3$$

In this case the brackets are there just to help you to understand the different steps. We could have left them out completely.

Activity 2.7

Adding expressions

Add the following pairs of expressions:

1. $4a\text{--}3b$ and $2a+2b$

2. $4a\text{--}3b+c$ and $2a+2b\text{--}3c$

3. $4a\text{--}3b+c$ and $\text{--}2a+c$

4. $2a+2b\text{--}3c$ and $\text{--}2a+b+3c$

2.7 Subtracting expressions

Subtraction is a little bit more difficult.

Example 2.8

Subtract $2g+4$ from $3g-1$.

Let's see how this is done first:

$$(2g+4)-(3g-1) = (2g+4)-3g+1$$

$$= -g+5$$

The most important thing to notice in this calculation is that when you subtract the second expression you are subtracting each term of the second expression, so each term changes sign. In other words, $-(3g-1)$ becomes $-3g+1$. Now you try it yourself.

Activity 2.8

Subtracting expressions

Subtract the second expression from the first:

1. $4a-3b$ minus $2a+2b$

2. $4a-3b+c$ minus $2a+2b-3c$

3. $4a-3b+c$ minus $-2a+c$

4. $2a+2b-3c$ minus $-2a+b+3c$.

2.8 Dividing expressions

Finally, we can divide expressions. First remember that when you learned about fractions you found out how to put a fraction in its lowest terms. So:

$$\frac{2}{6}=\frac{1}{3}$$

To do this, we notice that on the left-hand side both the numbers 2 and 6 can be divided by 2. So, in effect, the 2s cancel each other:

$$\frac{2}{6}=\frac{1\times2}{3\times2}=\frac{1\times\cancel{2}}{3\times\cancel{2}}$$

$$=\frac{1}{3}$$

When dividing two algebraic expressions, you follow the same rules.

Example 2.9: Divide $6ab$ by $2a$.

You can divide the first expression and the second by 2 and a. So just as with fractions, we get:

$$\frac{6ab}{2a}=\frac{2a\times3b}{2a\times1}=\frac{\cancel{2a}\times3b}{\cancel{2a}\times1}$$

$$=\frac{3b}{1}=3b$$

Any number divided by 1 is itself, so $3b/1 = 3b$.

Activity 2.9

Dividing expressions

Divide the first expression by the second:

1. $4xy$ divided by $2y$

2. $4xy$ divided by $2x$

3. $4xy$ divided by $4x$

4. $4xy$ divided by xy

2.9 More about equations

Now that you have recapped on the various operations you can do with algebra and have done a few activities to refresh your memory, let's move on to the more complicated matter of solving equations. To begin with, let's discuss some important properties of equations.

An equation is just like a set of scales that are perfectly balanced. No matter what the expressions are on the left-hand side and the right-hand side of an equation, they represent the same number. You already know this, of course, because you found this out at the beginning of this chapter. In fact, this is exactly what the equals sign means.

For us this implies that if you do something to one side of an equation you *must* do the same thing to the other side, otherwise you will get an unbalanced set of scales.

This is the most important thing that you need to understand in order to solve equations.

2.10 Solving equations

Let's consider the simple equation:

$$a\text{-}4 = 2$$

You can probably tell the answer to this equation straight away. However, we will use it to illustrate the way in which we are going to solve equations in general.

The basic idea is to manipulate this equation so that on the left-hand side we only have a, and on the right-hand side we only have a number. To do this we need to systematically add, subtract, multiply or divide both sides of the equation.

In our equation we have a-4 on the left-hand side, but we want to have just a. To get from a-4 to a we just add 4. But, if we just do this to the left-hand side, we then get an unbalanced set of scales.

So we must add 4 to both sides of the equation, so we get:

$$a\text{-}4\text{+}4 = 2\text{+}4$$

or simply:

$$a = 6$$

Once our equation is written like this we have solved it, because this equation says that a is the number 6.

To summarise, we solve equations by cleverly adding, subtracting, multiplying or dividing both sides of an equation to end up with a letter on one side and a number on the other, remembering that what we do to one side of an equation we must also do to the other.

Example 2.10

Let's try to solve the equation:

$$4x+3 = 11$$

To solve this equation we need to manipulate it until it has x on the left-hand side and a number on the right.

The first thing we will do is to subtract 3 from both sides of the equation so that we get:

$$4x+3-3 = 11-3$$

or $4x = 8$

Now to get x by itself on the left-hand side we need to divide both sides of the equation by 4, remembering how we do that from the first chapter:

$$\frac{4x}{4} = \frac{8}{4}$$

$$\frac{4x}{4} = 2$$

$$x = 2$$

The last equation gives us the solution, x is 2. Now you try solving some equations in the following activities.

Activity 2.10

Solving equations 1

Find the solution to the equation $6g-3 = 15$.

Activity 2.11

Solving equations 2

Find the solution to the equation $3g+8 = 11$.

Sometimes you can have letters on both sides of an equation, for example:

$$6y-1 = 4y+3$$

To solve an equation like this, you use exactly the same method you have been using so far. Remember that the idea is to gather together letters on one side of the equation and numbers on the other.

Example 2.11

Solve $6y-1 = 4y+3$

First let us decide to gather the terms involving y on the left-hand side and terms involving only numbers on the right-hand side.

We add 1 to both sides:

$$6y-1+1 = 4y+3+1$$

so $6y = 4y+4$

Now subtract $4y$ from both sides:

$$6y-4y = 4y-4y+4$$

$$2y = 4$$

Now divide both sides by 2:

$$\frac{2y}{2} = \frac{4}{2} = 2$$

So, finally we have:

$$y = 2$$

Example 2.12

Solve $2 = \dfrac{3}{x}+1$

Again we gather terms involving x on the left-hand side and terms involving only numbers on the right-hand side.

Subtract 1 from both sides:

$$2-1 = \frac{3}{x}+1-1 \text{ so } 1 = \frac{3}{x}$$

Now multiply both sides by x so that x doesn't appear on the right-hand side:

$$1 \times x = \frac{3}{x} \times x \text{ so } x = 3$$

2.11 Formulas

You found out in the last two sections that equations can be thought of as pieces of information written using algebra. For example, the sentence 'When I retire in 40 years, I will be three times as old as I am now' was the same as the equation $a+40 = 3a$. Sometimes, however, you can use equations to tell you information about how one quantity changes when another does. For example, if the temperature is measured both in Celsius and in Fahrenheit and given by the letters C and F then the following equation relates the two quantities:

$$C = \frac{5}{9}(F-32)$$

When an equation is given like this, so that it relates one quantity to another, it is called a *formula*.

Example 2.13

What is the temperature in Celsius when it is 90 degrees Fahrenheit?

Here $F=90$ and so putting this value into the formula we get:

$$C = \frac{5}{9}(90-32)$$

$$= \frac{5}{9} \times 58$$

$$= 32\frac{2}{9}$$

Formulas are very useful indeed and they give you a lot of information. In fact you should think of a formula as providing you with additional information. For example, the formula above tells you that if you measure the temperature in Fahrenheit then you can tell what the temperature is in Celsius. In other words, you don't have to measure the temperature twice (once in Fahrenheit and once in Celsius) since you can work one out from another.

The letters in a formula are slightly different to the letters in an equation. In an equation the letter always represented a certain number that we did not yet know. In a formula, the letters do not represent specific numbers, but represent a varying number. It is for this reason that the letters in a formula are called *variables*.

The formula above shows us that if you tell me a temperature in Fahrenheit then I can work out the temperature in Celsius. In other words, if you give me a value for F, then I can work out the corresponding value for C. When a formula relates the two variables, F and C, like this, F is called the *independent variable* and C is called the *dependent variable*, since the value of C depends on what the value of F is. Also, we say that C is the *subject* of the formula.

Example 2.14

What is the temperature in Celsius when it is −10 degrees Fahrenheit?

Here $F=-10$, so putting this value into the formula we get:

$$C = \frac{5}{9}(-10-32) = \frac{5}{9}(-42) = -23\frac{1}{3}$$

2.12 Changing the subject of a formula

In the last section we saw that we could relate the temperature in Celsius to the temperature in Fahrenheit using the formula:

$$C = \frac{5}{9}(F-32)$$

Suppose now that we are given the temperature in Celsius, C, and we need to find out the temperature in Fahrenheit, F. How would you do this?

Example 2.15

1. What is the temperature in Fahrenheit when it is 40° Celsius?

    By the formula above we have:

    $$40 = \frac{5}{9}(F - 32)$$

    and so we have an equation in F that we can solve:

    $$40 = \frac{5}{9}(F - 32)$$

    $$40 \times \frac{9}{5} = (F - 32)$$

    $$72 = F - 32$$

    $$104 = F$$

    So, it is 104° Fahrenheit when it is 40° Celsius.

2. What is the temperature in Fahrenheit when it is −10 degrees Celsius?
    By the formula above we have:

    $$-10 = \frac{5}{9}(F - 32)$$

    and solving this for F as before we get:

    $$(-10) \times \frac{9}{5} = F - 32 \text{ so } -18 = F - 32 \text{ so } -18 + 32 = F \text{ so } 14 = F.$$

    So, it is 14° Fahrenheit when it is −10° Celsius.

Suppose now that you need to know the temperature in Fahrenheit when, in Celsius, it is 50, 60, 70, . . . ! It would soon become rather cumbersome to keep solving equations like we did in the examples. The problem is that in the formula we have above, C is the dependent variable or the subject of the formula. It would be easier now if we had F as the subject of the formula since then we could just work out the temperature by putting the value of C in the formula.

The process of making F the dependent variable and C the independent variable is called *changing the subject* of the formula. It turns out that changing the subject is exactly like solving an equation as we did in section 1.10.

Example 2.16

Make F the subject of the formula:

$$C = \frac{5}{9}(F - 32)$$

To do this we need to 'solve' this formula so that on one side we have F by itself and on the other side we have a relationship involving C. We will proceed just as we did when we were solving equations.

$$C = \frac{5}{9}(F-32)$$

Multiplying by 9/5:

$$\frac{9}{5}C = \frac{9}{5} \times \frac{5}{9}(F-32)$$

$$\frac{9}{5}C = F-32$$

Adding 32 to both sides:

$$\frac{9}{5}C+32 = F-32+32$$

$$\frac{9}{5}C+32 = F$$

The last equation gives us the formula for F in terms of C. This means that the subject of the formula is F.

It is now easy to work out the temperature in Fahrenheit, given the temperature in Celsius.

Example 2.17

What is the temperature in Fahrenheit when it is 40 degrees Celsius? Here $C = 40$ so:

$$F = \frac{9}{5} \times 40 + 32 = 9 \times 8 + 32 = 104$$

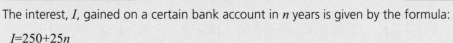

Activity 2.12

Changing the subject 1

The interest, I, gained on a certain bank account in n years is given by the formula:

$I=250+25n$

Make n the subject of this formula and find n when $I=500$.

Activity 2.13

Changing the subject 2

The cost c in pence of posting n copies of a book is given by the formula:

$c=100+23n$

Make n the subject of this formula. If I spent 192 pence, how many copies did I send?

Activity 2.14

Changing the subject 3

The length of time, t seconds, for a ball of weight w kilograms to stop bouncing after falling 1 metre is given by the formula:

$$t = \frac{4n}{w}$$

Make w the subject of the formula. Calculate the weight of the ball needed to ensure that it stops bouncing after 6 seconds.

2.13 Turning problems into equations

By now you have found out how to solve equations by algebraic manipulation. You have also found out how different quantities can be related by a formula, and how to change the subject of that formula. These are all very useful tools that you need to solve problems in algebra. However, so far all the examples and activities started out by giving you an equation or a formula and asking you to solve it or otherwise manipulate it, using everything you have learned in this chapter so far.

In reality, however, many problems are not stated in such a simple way, and you have to be able to change a written problem into an equation in algebra. A good example of this was the example that asked you to work out someone's age. In this example you were not given an equation, only the sentence 'When I retire in 40 years, I will be three times as old as I am now.'

In order to solve this, we first had to rewrite this sentence as the equation:

$a+40 = 3a.$

Then we solved this equation to find out the age.

Most problems that you will encounter will be given like this. The first hurdle is, therefore, to be able to write such a problem as an equation.

The unknown quantity

The first step in solving any problem is to decide what the unknown quantity is that you are looking for and representing it by a letter. This is usually quite easy to do.

Example 2.18

My electricity bill per quarter is calculated as £40 plus 25 pence per unit used. If I get a bill for £55, how many units of electricity have I used?

In this example the unknown quantity is evident from the final part of the question, 'How many units of electricity have I used?' The quantity that we are looking for is the number of units. We will represent this quantity by the letter u.

Finding the equation

The second step, and usually the most difficult, is to formulate the problem as an equation.

Example 2.19

In the previous example we need to find the number of units of electricity used. However, to do this we are given information about my electricity bill. The first sentence tells us how the bill is calculated from the number of units and the second tells us the total sum of the bill.

So, the first sentence tells us that the bill is '£40 plus 25 pence per unit used'. So, if we use u for the number of units we have that the bill is $40+0.25u$.

Notice that we have $0.25u$ since 25 pence in pounds is the same as £0.25. For a formula to work, you must use the same units throughout.

The second sentence tells us that 'I get a bill for £55'. So we have the equation $40+0.25u=55$.

Solving the equation

In the beginning of this chapter you learned that algebra can be used to simplify problems in order to solve them. So, the final step to solving the problem is always to solve an equation.

Example 2.20

We are now able to solve how many units of electricity I have used this quarter. In order to do this we have to solve the equation $40+0.25u = 55$:

$$40+0.25u = 55$$

Subtract 40:

$$40+0.25u-40 = 55-40$$

$$0.25u = 15$$

Divide by 0.25:

$$\frac{0.25u}{0.25} = \frac{15}{0.25}$$

$$u = 60$$

Hence, I used 60 units of electricity this quarter.

Example 2.21

When travelling by car, it takes me 10 minutes to walk to my car and start it up; I then travel at an average speed of 30 miles per hour. If it takes me 45 minutes to get to work, how far is it away?

In this problem, the unknown is the distance it is to work. We will call this d.

The first sentence tells us how long it takes me to travel d miles: it takes 10 minutes (1/6 of an hour) and d/30 hours. So the time it takes me to travel d miles is:

$$\frac{1}{6} + \frac{d}{30}$$

But the second sentence tells us that it takes me ¾ of an hour to travel to work, so, we have the equation:

$$\frac{1}{6}+\frac{d}{30}=\frac{3}{4}$$

Now, if we now solve this for d we get:

Subtract $\frac{1}{6}$:

$$\frac{d}{30}+\frac{1}{6}-\frac{1}{6}=\frac{3}{4}-\frac{1}{6}$$

Multiply by 30:

$$\frac{d}{30}=\frac{7}{12}$$

$$30 \times \frac{d}{30} = 30 \times \frac{7}{12}$$

$$d = 17\frac{1}{2}$$

So it is 17.5 miles to work.

Example 2.22

Producing a batch of CDs costs £10 to set up the press and then 32p for each CD produced. If I have £25, how many CDs can I produce (where £1 = 100p)?

The unknown is the number of CDs. We will call this n. The first sentence says that the cost is 1,000 pence plus $32n$ pence. The second sentence says that if I spend all my money then the cost is 2,500 pence. So we have the equation:

$1000+32n=2500$

Solving this we get:

$32n=2500-1000=1500$ so $\quad n = \dfrac{1500}{32} = 46.875$

However, since I obviously can't produce part of a CD in fact I'll be able to produce 46 CDs.

Example 2.23

Ali buys a cake and three drinks, which cost him 96p. If a cake costs 8p more than a drink, how much does a drink cost? How much does a cake cost?

Here we have two unknowns: the price of a cake, which we will call c, and the price of a drink, which we will call d.

The first sentence says that the cost of a cake (c) plus the cost of three drinks ($3d$) is 96p. As an equation this is:

$c+3d=96$

The second sentence says that the cost of a cake (c) is 8p more than the cost of a drink (d). So c must be equal to d plus 8, so as an equation:

$c=d+8$

Since both the equations must be true we can put this value of c from the second equation into the first equation. This gives us:

$(d+8)+3d=96$

Now if we collect together the like terms we get:

$4d+8=96$ so $4d=96-8=88$ so $\dfrac{88}{4} = 22$

So a drink costs 22p. Using the second equation:

$c=22+8=30$

So a cake costs 30p.

2.14 Summary

In this chapter we saw how to use equations to state relationships between quantities, and explore these relationships. We saw how to manipulate equations using the algebra techniques from chapter 1, and how to solve linear equations to find the values we need. These ideas will be used throughout the rest of the module, but particularly in chapter 4 on proportions, percentages and ratios and in the chapters on statistics.

2.15 Review questions

Question 2.1: Equations Solve the following equations:

1. $4t = 8$

2. $6h = 3h+9$

3. $8k+4 = 10k+2$

Question 2.2: Subjects What is the subject in the following formulas?

1. $C = \frac{5}{9}(F-32)$

2. $y = 4x-3$

3. $R = (1+i)^{20}$

Question 2.3: Formula If the speed of a car increases by 3 kilometres per hour every second, how long will it take for the car to accelerate from 60kph to 90kph?

Question 2.4: Finding my age

When I retire in 40 years, I will be three times as old as I am now. How old am I?

Write down the equation we had that represented this sentence. Solve this equation and find out how old I am.

Question 2.5: Dependent and independent variables

Write down which variable is the dependent variable and which is the independent variable in the following formulas:

Formula	Dependent variable	Independent variable
$a = 2v+3$		
$y = 5(x-2)$		
$L = 2\pi/G$		

Question 2.6: Subject of a formula

Write down which variable is the subject of the following formulas.

Formula	Subject
$l = 2^x$	
$e = mc^2$	
$v = (a\text{-}3)$	

Question 2.7: Solving equations 1

An engineer is building a bridge over a river. She calculates that the stress level of the bridge, when nothing is going over it, is 2.5 tonnes per square metre. In addition to this, she calculates that vehicles travelling over the bridge add 0.2 tonnes per square metre for every tonne of weight. If the bridge can take, at most, 10 tonnes per square metre of stress, how much weight can the bridge carry?

Question 2.8: Solving equations 2

A computer programmer works out that his server uses 10% of his computer's memory plus 3% per user logged in. How many users can his server cope with?

Question 2.9: Solving equations 3

At a certain temperature, both the readings on the Celsius and Fahrenheit scales are the same. What is this temperature?

2.16 Answers to review questions

Question 2.1: Equations

1. $4t = 8$. Divide by 4:

   $t = 2$

2. $6h = 3h\text{+}9$. Subtract $3h$:

   $3h = 9$. Divide by 3:

   $h = 3$

3. $8k\text{+}4 = 10k\text{+}2$. Subtract $8k$:

   $4 = 2k\text{+}2$. Subtract 2:

   $2 = 2k$. Divide by 2:

   $2 = k$.

Question 2.2: Subjects

1. C
2. y
3. R.

Question 2.3: Formula

Let t be the time in seconds, then we know that $3t$ is the total difference in speed. So, since the difference in speed is 90-60 = 30kph we have the equation:

$$3t = 30$$

If we divide both sides by 3, we see that $t = 10$ and so it takes 10 seconds for the car to accelerate from 60kph to 90kph.

Question 2.4: Finding my age

The equation was $3a = a+40$. Subtract a:

$3a-a=a+40-a$

$2a=40$. Divide by 2:

$\frac{2a}{2}=\frac{40}{2}$

$a=20$

I am 20 years old.

Question 2.5: Dependant and independant variables

Formula	Dependent variable	Independent variable
$a=2v+3$	a	v
$y=5(x-2)$	y	x
$L=2\pi/G$	L	G

Question 2.6: Subject of a formula

Formula	Subject
$l=2^x$	l
$e=mc^2$	e
$v=(a-3)$	v

Question 2.7: Solving equations 1

Here the unknown is the weight, say w tonnes.

The engineer works out that the stress is 2.5 tonnes per square metre plus 0.2 tonnes per square metre for every additional tonne. So the total stress is $25+0.2w$ tonnes per square metre.

The bridge can only take 10 tonnes per square metre of stress, so we need to solve the equation above. Subtract 2.5:

$2.5+0.2w-2.5 = 10-2.5$

$0.2w = 7.5$. Divide by 0.2:

$$\frac{0.2w}{0.2} = \frac{7.5}{0.2}$$

$w = 37.5$ tonnes

So, the maximum weight that the bridge can carry is 37.5 tonnes.

Question 2.8: Solving equations 2

The unknown in this example is the number of users, u. The programmer works out that 10% of the computer's memory is used by the server and 3% per user. So the total memory used is $10+3u$ per cent. Of course, the server cannot use more than 100% of the memory and so we need to solve the equation:

$10+3u = 100$. Subtract 10

$3u=90$. Divide by 3:

$$\frac{3u}{3} = \frac{90}{3}$$

$u=30$

So the server can have, at most, 30 users at one time.

Question 2.9: Solving equations 3

This activity requires a little more thought. The formula that links Celsius and Fahrenheit is $C=\frac{5}{9}(F-32)$ as we have already seen. Now, the question asks for the value when the two scales read the same, that is, when does $X^oF = X^oC$?

To solve this we simply write C instead of F in the above formula and solve.

$C=\frac{5}{9}(C-32)$ Multiply by 9/5:

$\frac{9}{5}C=\frac{9}{5}\frac{5}{9}(C-32)$

$\frac{9}{5}C=C-32$ Subtract C:

$\frac{9}{5}C-C=C-32-C$

$\frac{4}{5}C=-32$ Multiply by 5/4:

$\frac{5}{4}\frac{4}{5}C=\frac{5}{4}\times(-32)$

$C=-40$

So -40 Fahrenheit is the same as -40 Celsius.

2.17 Feedback on activities

Activity 2.1: Bracket basics

2x4+3 = 11

2x(4+3) = 2x7=14

Activity 2.3: More on brackets

The first two expressions work out to be $3x$-2. The third expression is different.

Activity 2.5: Equations as sentences

Twice a is the same number as a plus 4.

Three times a is the same number as two times a minus 3.

Activity 2.6: Multiplying variables

1. $4x \times 3y = 12xy$
2. $3x \times 4y = 12xy$
3. $3y \times 4x = 12xy$
4. $4x \times 3x = 12x^2$

Activity 2.7: Adding expressions

1. $(4a–3b) + (2a+2b) = 2a–b$
2. $(4a–3b+c) + (2a+2b–3c) = 2a–b–2c$
3. $(4a–3b+c) + (–2a+c) = 2a–3b+2c$
4. $(2a+2b–3c) + (–2a+b+3c) = 3b$

Activity 2.8: Subtracting expressions

1. $(4a–3b) – (2a+2b) = (4a–3b)–2a–2b = 2a–5b$
2. $(4a–3b+c) – (2a+2b–3c) = (4a–3b+c)–2a–2b+3c = 2a–b+4c$
3. $(4a–3b+c) – (–2a+c) = (4a–3b+c)+2a–c = 6a–3b$
4. $(2a+2b–3c) – (–2a+b+3c) = (2a+2b–3c)+2a–b–3c = 4a+b–6c$

Activity 2.9: Building expressions

1. $\dfrac{4xy}{2y} = 2x$

2. $\dfrac{4xy}{2x} = 2y$

3. $\dfrac{4xy}{4x} = y$

4. $\dfrac{4x}{xy} = 4$

Activity 2.10: Solving equations 1

$6g-3=15$.   Add 3 to both sides:

$6g-3+3=15+3$

$6g=18$.   Divide by 6:

$$\frac{6g}{6}=\frac{18}{6}$$

$$\frac{6g}{6}=3$$

$g=3$

Activity 2.11: Solving equations 2

$3g+8=11$.   Subtract 8:

$3g+8-8=11-8$

$3g=3$.   Divide by 3:

$$\frac{3g}{3}=\frac{3}{3}$$
$g=1$

Activity 2.12: Changing the subject 1

Subtract 250:

$I-250=250+25n-250$

$I-250=25n$.   Divide by 25:

$$\frac{I-250}{25}=\frac{25n}{25}=n$$

So when $I = 500$:

$$n=\frac{500-250}{25}=10\,years$$

Activity 2.13: Changing the subject 2

$c=100+23n$ so $c-100=23n$ so $\dfrac{c-100}{23}=n$.

So if I spend 192 pence then $c = 192$ so

$$n=\frac{192-100}{23}=\frac{92}{23}=4$$

so I sent 4 books.

Activity 2.14: Changing the subject 3

Multiply by w:

$$tw = \frac{4\pi}{w} w$$

$tw = 4\pi$ Divide by t

$$\frac{tw}{t} = \frac{4\pi}{t}$$

$$w = \frac{4\pi}{t}$$

So to make sure the ball stops bouncing after 6 seconds we let $t = 6$.

Then we have:

$$w = \frac{4\pi}{6} = \frac{2\pi}{3} \approx 2.09\text{kg}$$

CHAPTER 3

More on equations

OVERVIEW

In this chapter we'll look at linear and quadratic equations. These are equations involving x, and equations involving terms such as x^2 (*xxx*) as well as simply x. They're a bit more complicated than the equations we looked at in chapter 2, but they are needed to deal with some real-world problems. We'll see how to represent linear and quadratic functions pictorially, using graphs. We'll also explore various ways of finding solutions of quadratic equations.

We'll also look at cubic equations and other polynomials, and exponential and logarithmic functions. Again, these are useful in dealing with some real-world problems.

Learning outcomes On completion of this chapter, you should be able to:
• Plot a graph of a linear, quadratic or other polynomial function
• Solve a linear equation
• Rearrange a quadratic equation into standard form
• Find out how many solutions you can get from a quadratic equation or other polynomial
• Solve a quadratic equation by estimation and by using the quadratic formula
• Explain what is meant by exponential and logarithmic functions, and plot their graphs.

You should have completed chapter 2 of this book before starting this chapter.

3.1 Plotting a graph

In this chapter we'll be exploring **functions** that describe a relationship between two variables (usually x and y). Given a value of x, we can calculate a corresponding value of y. A useful way to look at these pairs of values is using a **graph**.

Figure 3.1: x and y functions viewed as a graph

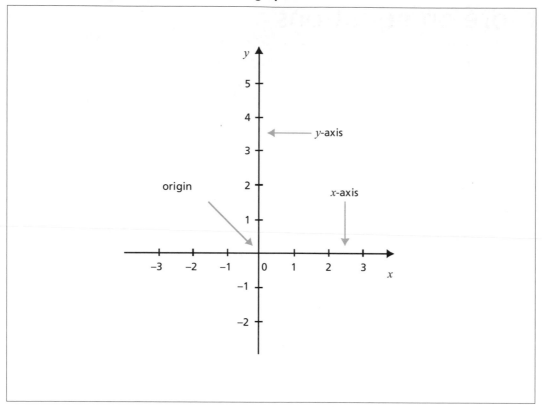

To draw a graph we start with a set of **axes**: a horizontal line called the x-**axis** that represents x, and a vertical line called the **y-axis** that represents y. The distances along each axis are labelled, as shown. Usually we put a label at every unit, but sometimes we include only alternate ones, for example to stop the diagram getting too cluttered. The point where the axes cross is called the **origin**.

Suppose we have the pair of values $x = 2$ and $y = 4$, for example. To represent this point on the graph we start from the origin. We go 2 units horizontally, then 4 units vertically, and draw a dot to represent the point.

Figure 3.2: Graph co-ordinates

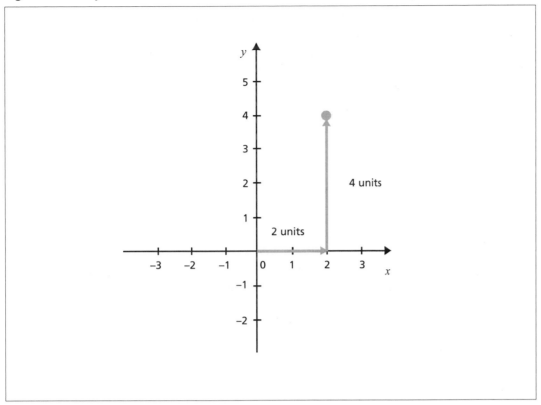

We sometimes write a pair of *x* and *y* values as (*x, y*). If we have a point on the graph, then the (*x, y*) values that tell us where it is are called the **co-ordinates** of the point. Thus, the co-ordinates of the point in the above example are (2,4).

Activity 3.1

Plotting points

Draw a set of axes with the *x* values going from –5 to 5 and the *y* values going from –3 to 3. Plot the following points – this means represent them on the graph by drawing a dot in the correct place:

1. (2,3)

2. (3,2)

3. (-1,2)

4. (-2,-3).

Label each point.

3.2 Linear equations

An equation defines a relationship between two or more variables. For example, in section 2.11 we defined a relationship between temperature in Celsius and temperature in Fahrenheit. We can represent the relationship on a graph by plotting points to indicate corresponding values of the variables. Often these points will join up to form a straight line or curve representing the relationship.

In chapter 2 we looked at linear equations, which means equations involving variables such as x and y but not powers such as x^2 or x^3 or terms such as xy. If we plot a graph of a linear equation by marking points where the x and y values are related by the equation, then we get a straight line.

Example 3.1

Suppose we have the equation:

$y = x + 2$

If $x = 0$ then we can put this into the right-hand side and find that the corresponding value of y is 2. Thus we have a pair of values satisfying the equation, and we can represent them by the point (0, 2). If $x = 1$, then the corresponding value of y is 3, and we can represent these values by the point (1, 3). If $x = -1$ then the corresponding value of y is 1, represented by the point (-1, 1), and so on. If we continue, marking *all* the points representing possible pairs of values satisfying this equation, then we get:

Figure 3.3: The function $y = x + 2$

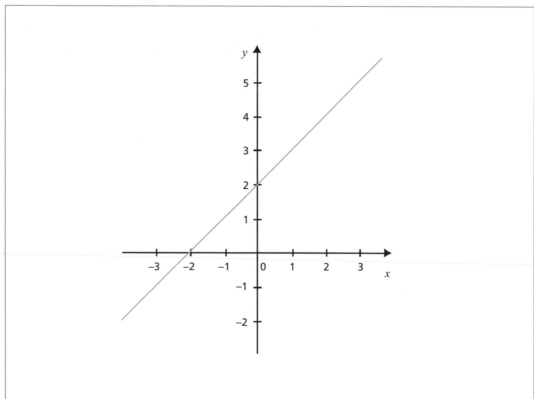

It turns out that the graph of a linear equation is always a straight line. This means that in fact to plot the line we need find only two points, and then join these points up with a straight line. The two points can be anywhere on the line, but it's sensible to pick one value of x near the left-hand end of the region we're interested in and the other near the right-hand end.

For example, to plot the line in figure 3.3 we might pick $x = -3$ which gives $y = -1$ and hence the point $(-1, -3)$, and $x = 3$ which gives $y = 5$ and hence the point $(3, 5)$. We can plot the points $(-1, -3)$ and $(3, 5)$ on the graph and join them up with a straight line to get the line shown.

Activity 3.2

Plotting a linear equation

Draw a set of axes with the x values going from -1 to 3 and the y values going from -4 to 4. Draw the line representing the equation

$$y = 2x - 2.$$

Hint: pick two values of x. For each one, work out the value of y, and plot the corresponding point on the graph. Join up the two points with a straight line.

We can change a linear equation into several different forms, as we did in chapter 2. One standard form is to write it as:

$$y = mx + c$$

where x and y are the two variables, and m and c are constants (often just numbers). The equation in example 3.1 was in this form, with $m = 1$ and $c = 2$. The equation in activity 3.2 was also in this form, with $m = 2$ and $c = -2$.

If we put an equation into this form then the constant m tells us the **slope** or **gradient** of the straight line. A positive slope means that the line goes from bottom-left to top-right (upwards), and the larger the slope the steeper the line. A negative slope means that the line goes from top-left to bottom-right (downwards), and the more negative the slope the steeper the line. A slope of 0 means that the line is horizontal. For example, the equation in example 3.1 has $m = 1$ so the line has positive slope. The equation in activity 3.2 has $m = 2$ so it also has positive slope, and its line is steeper.

The constant c tells us where the line crosses the y-axis. This point is sometimes called the **intercept**. For example, the equation in example 3.1 has $c = 2$, and its line crosses the y-axis at $y = 2$. The equation in activity 3.2 has $c = -2$, and its line crosses the y-axis at $y = -2$.

Some people use $y = a + bx$ instead of $y = mx + c$, so here a is the intercept and b is the slope.

Another way to think about the slope is that it tells us how y changes as x changes. If x increases by 1 unit then y changes by m units. So if $m = 2$ then a change of 1 in x causes a change of 2 in y. If $m = -3$, then a change of 1 in x causes a change of -3 in y, and so on.

Activity 3.3

Slopes and intercepts

Consider the equation

$$y = 3 - 2x$$

What is the value of m here? Do you expect the line to slope upwards or downwards. How steep will it be?

What is the value of c here? Where do you expect the line to cross the y-axis?

Plot the line on a graph (by picking two values of x as usual) and see whether your predictions were correct.

Another standard form for a linear equation is to write it in the form:

$$k = ax + by$$

where as usual x and y are the two variables and a, b and k are constants. This looks different from the $y = mx + c$ form. However, from chapter 2 we know how to make y the subject of this formula:

$$k = ax + by \text{ so } k - ax = by \text{ so } \frac{k}{b} - \frac{ax}{b} = y \text{ so } y = -\frac{a}{b}x + \frac{k}{b}$$

and the right-hand side is now in the form $mx + c$. So these two ways of writing linear equations are really just different arrangements of the same thing.

Example 3.2

Suppose we have the equation

$$6 = x + 2y$$

Rearranging this we get:

$$6 = x + 2y \text{ so } 6 - x = 2y \text{ so } \frac{6}{2} - \frac{x}{2} = y \text{ so } y = 3 - \frac{x}{2}$$

Activity 3.4

Rearranging linear equations

Rearrange the following linear equations into the form $y = mx + c$:

1. $10 = 3x + 2y$

2. $1 = -x - 3y$

3. $-3 = 3y - 4x$

We saw that we can plot a graph of a linear equation by finding two points on the line and joining them up. Up to now we've just picked two convenient values of x to get the two points. However, if the equation is in the form $k = ax + by$ then it's often convenient to use the point where $x = 0$ and the point where $y = 0$, since these points are easy to find. Also, they lie on the axes so they're easy to plot accurately.

Example 3.3

Consider the equation $6 = x + 2y$.

If $x = 0$ then we have:

$$6 = 0 + 2y = 2y \text{ so } y = \frac{6}{2} = 3$$

so the point (0, 3) lies on the line.

If $y = 0$ then we have:

$$6 = x + 2(0) = x$$

so the point (6, 0) lies on the line (be careful to get the x and y values the right way round!). So we can plot the points (0, 3) and (6, 0), and then join them up to get a graph of the equation:

Figure 3.4: Graph of the function $6 = x + 2y$

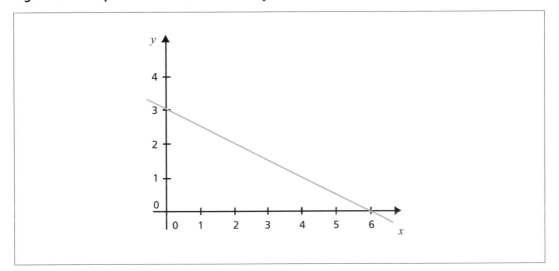

Activity 3.5

Plotting a linear equation

Plot a graph of the function $10 = 3x + 2y$ by plotting a point on the x-axis and a point on y-axis.

3.3 Quadratic functions

In the previous section we explored equations involving **linear functions**. In a linear function a variable may be multiplied by a constant, but we never have two variables multiplied together in the same term.

A **quadratic function** is one that also involves x^2. Remember that this means x multiplied by itself, or xxx. In this module we'll consider only quadratic functions that involve a single variable, usually x.

A **quadratic equation** is an equation involving quadratic functions. Usually the equation says that a given quadratic function is equal to 0, for example $x^2 = 0$.

Plotting a quadratic function

To see what a quadratic function looks like; we can plot it on a graph. Let's consider the simplest possible quadratic function:

$y=x^2$ (equation 1)

Given any x value, we can use the function to work out the corresponding y value. This gives us a point (x,y) that we can plot on a graph. By doing this for lots of different x values we can build up a picture of what the function looks like.

In fact, as with a linear function, since we could choose *any* x value there are infinitely many x points that we could plot. However, we expect that the points will lie on some nice, smooth curve. So, in practice, we can just pick a range of values of x, work out the corresponding y values and plot these points. We can then join these points up with a smooth curve to get a picture of the whole function, or part of the function.

So, consider the value $x = 2$. Using *equation 1*, the corresponding value of y is: $x^2 = 2^2 = 2 \times 2 = 4$.

You can do this calculation on a calculator using the 'squared' button, which will probably be labelled 'x^2'. For example, to calculate 2^2, press:

and you should get 4. We can find y for various other values of x similarly:

x	0	0.5	1	2	3
y	0	0.25	1	4	9

(Notice that when $x = 0.5$ we have $y = 0.5 \times 0.5 = 0.25$.)

What about negative values of x? If $x = -1$, then:

$y = x^2 = (-1)^2 = (-1) \times (-1) = 1$

In fact, if we square any negative number then we get a positive number, because 'minus times minus equals plus'. So, we can extend our table of values.

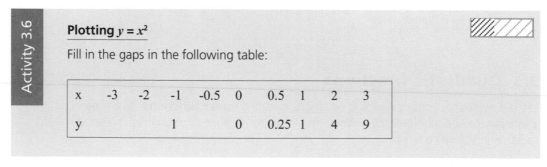

Activity 3.6

Plotting $y = x^2$

Fill in the gaps in the following table:

x	-3	-2	-1	-0.5	0	0.5	1	2	3
y			1		0	0.25	1	4	9

We can then plot these points on a graph, as we did in activity 3.1. If we plot each point as a dot, then we get this:

Figure 3.5: Plotted points of $y = x^2$

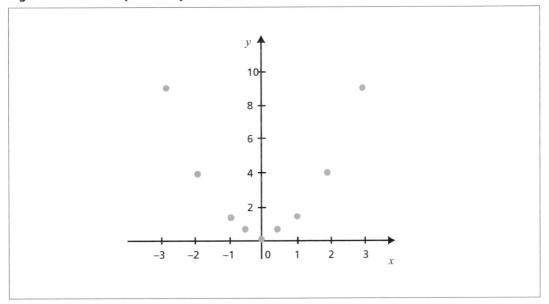

Finally, we can join the points up with a smooth curve like this:

Figure 3.6: Graph of the function $y = x^2$

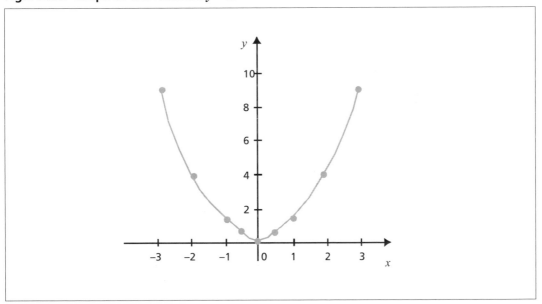

Thus we have a graph of the function $y = x^2$.

This shape of curve is called a **parabola**. A parabola is horizontal right at the centre, but rises more and more steeply as x moves away from the centre in either direction. It is **symmetrical** about a vertical line through the centre. This particular parabola has the centre at the origin, so it's symmetrical about the y-axis.

A curve like this that curves upwards from the middle is called **convex**.

3.4 General quadratic functions

In general a quadratic function is one of the form: $y = ax^2 + bx + c$.

Here a, b and c are **constants**, which means they have a fixed value. Often, they're simply numbers.

Thus, the function consists of a constant a times x^2, plus a second constant b times x, plus a third constant c. The constant a is called the **coefficient** of x^2, and so on.

If we plot a graph of any quadratic function, we will get a parabola similar to the one we drew for $y=x^2$, although the centre may be in a different place and it may be stretched or squashed.

Figure 3.7: Example 1

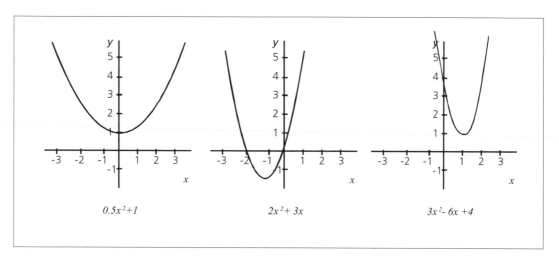

$0.5x^2+1$ $\qquad$ $2x^2+3x$ $\qquad$ $3x^2-6x+4$

Real-world examples of convex parabolas

1. A suspension bridge has two towers with cables hanging between them and the road suspended from these by smaller cables. Most suspension bridges are designed so that the main cables form a parabola.

Figure 3.8: Bridge example of convex parabola

2. If you shine a torch at an angle to a wall, the edge of the patch of light can be a parabola (but it depends on the angle).

What happens if *a* < 0?

If *a*, the coefficient of x^2, is negative, the parabola will be the other way up.

Figure 3.9: Example 2

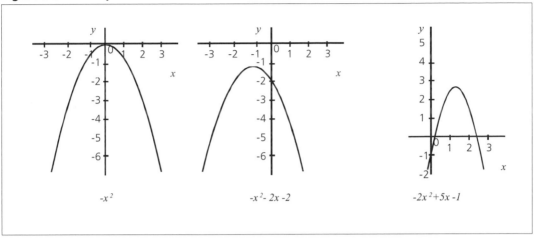

In these cases the parabola is horizontal right at the centre, but falls more and more steeply as *x* moves away from the centre in either direction. As before, it is symmetrical about a vertical line through the centre.

A curve like this that curves downwards from the centre is called **concave**.

Real-world examples of concave parabolas

1. If you throw a ball, the path that it follows in the air is a parabola.

Figure 3.10: Example of concave parabola

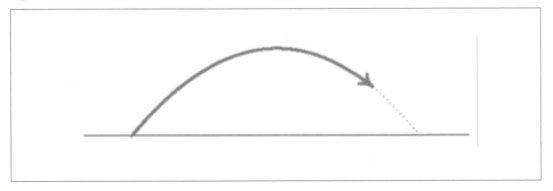

2. An arched bridge has two piers with an arch between them supporting the road. Most arched bridges are designed so that the arch forms a parabola.

Figure 3.11: Bridge example of concave parabola

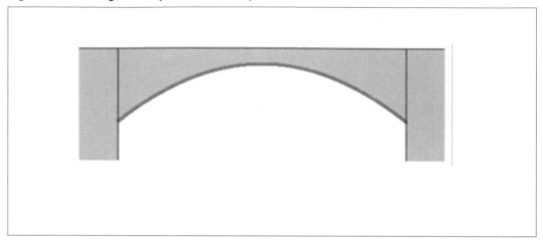

Plotting quadratic functions

Plot your own graphs of some of the quadratic functions in example 1 and example 2. You'll need to choose a range of values of x, then for each one calculate the corresponding y value. Plot these pairs of values on a set of axes, then join them up with a smooth curve to see the function. Check that your graphs are similar to the ones here.

3.5 The standard form of a quadratic equation

To make it easier to deal with a quadratic equation, it's helpful to put it into the **standard form**:

$$ax^2+bx+c = 0$$

The point is that we have a single term involving x^2, then a single term involving x, then a single constant term.

We can do this by collecting terms, and moving them from one side of the equation to the other, as we did in chapter 2.

Sometimes:

- We just have to put the terms in a different order:
  $3+x^2-2x = 0$ into $x^2-2x+3 = 0$

- We have to move a term from one side of the equation to the other. If we have:
  $2x^2+3 = 2x$ we can subtract $2x$ from both sides to get:
  $2x^2-2x+3 = 0$

- We need to collect terms together. If we have:
  $x^2+2x+2x^2-3x+5 = 0$ we can combine the x^2 and the $2x^2$ to make $3x^2$, and we can combine the $2x$ and the $-3x$ to make $-x$, giving us the equation:
  $3x^2-x+5 = 0$.

The signs of the coefficients

Once we have an equation in the standard form:

$$ax^2+bx+c = 0$$

then, when we're writing down the values of the coefficients a, b and c, it's important to get the signs correct (in other words, whether they're positive or negative).

For example, in the equation:

$$x^2-2x+3 = 0$$

the value of b is -2, not just 2.

This is particularly important when a is negative. In the equation:

$$-3x^2+x = 0$$

it's easy to forget the minus sign at the start, but here the value of a is -3, not just 3.

Activity 3.8

Quadratics in standard form

Put the following quadratic equations into standard form, and then write down the values of a, b and c:

1. $2 = 3x+5x^2$

2. $x-x^2+3x+1 = 3+x^2+4x-2$

3. $x(x+2) = 3x-4$. *Hint:* multiply out the bracket first

4. $-(x+1)(2-x) = 3-x-x^2$

5. $(1-x)(4+x) = (2x-1)(2x+2)$

The roots of a quadratic equation

Solving an equation means finding one or more values of x that make the equation true. Such values of x are called **solutions** or **roots** of the equation.

For example, consider the quadratic equation:

$$x^2-2x-3 = 0 \text{ (equation 2)}$$

This is a typical quadratic equation, with a quadratic function on one side and 0 on the other. To solve the equation we need to find values of x that make the equation true. In other words we're looking for values of x that make the quadratic function equal to 0.

Now, if we put the value $x = 3$ into this equation then the left-hand side is:

$$3^2-2\times3-3 = 9-6-3 = 0$$

which is equal to the right-hand side. Therefore $x = 3$ is a solution of this equation.

Activity 3.9

Checking a root

Check that $x = -1$ is also a solution of the equation above.

Hint: remember that squaring a negative number gives a positive number.

Thus equation 2 has at least two roots: $x = -1$ and $x = 3$.

The roots are where the parabola crosses the x-axis.

If a quadratic function is 0 at a certain value of x, when we plot the graph this point will lie on the x-axis. So, solving a quadratic equation corresponds to finding the place(s) where the parabola crosses the x-axis.

For example, suppose we take y to be the left-hand side of equation 2, i.e.

$$y = x^2 - 2x - 3$$

If we plot a graph of this function, we get:

Figure 3.12: Parabola crossing the x-axis

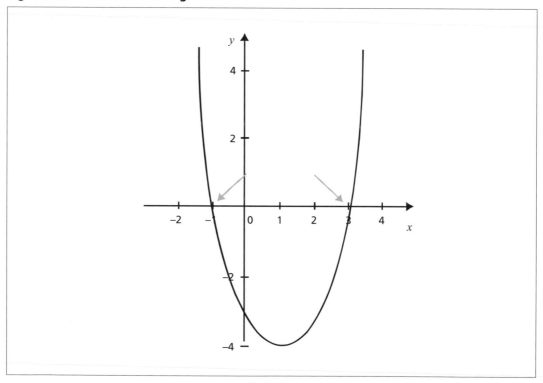

The parabola crosses the x-axis at the two points shown, which are the roots $x = -1$ and $x = 3$, as we'd expect.

The number of roots of a quadratic equation

How many roots can a quadratic equation have? To answer this question, we can ask how many times a parabola can cross the x-axis. We've seen that it might cross twice.

However, suppose that a, the coefficient of x^2, is positive so the parabola is convex. Then we saw earlier that the parabola rises more and more steeply as x moves away from the centre. So, once it's risen above the axis it can't curve back down to cross it again. A similar thing happens if $a < 0$: the parabola is concave, so once it has fallen below the axis it can't curve back up to cross it again.

Thus, a parabola can cross the x-axis, *at most,* twice. Thus, a quadratic equation *cannot* have more than two roots.

For example, we saw that equation 2 had two roots, $x = -1$ and $x = 3$. We now know that these are the only roots. There cannot be another value of x that makes equation 2 true.

Single roots

In some cases a quadratic equation has a single root. For example, consider the equation:

$x^2-4x+4 = 0$ (equation 3)

The graph of $y = x^2-4x+4$ looks like this:

Figure 3.13: Graph of $y =x^2-4x+4$

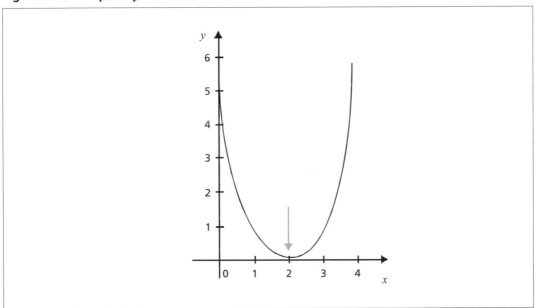

Here, as the value of x increases, the parabola comes down to the x-axis, but it just touches the axis once before it start to rise up again. The equation $x^2-4x+4 = 0$ therefore has only one root.

No roots

In other cases a quadratic equation has no roots at all. For example, consider the equation:

$x^2-2x+3 = 0$ (equation 4)

The graph of $y = x^2-2x+3$ looks like this:

Figure 3.14: Graph of $y = x^2-2x+3$

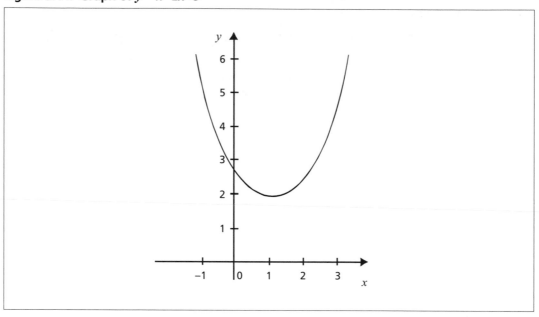

Here the parabola never crosses the x-axis at all. The equation $x^2-2x+3 = 0$ therefore has no roots. There is *no* value of x that makes this equation true.

3.6 Methods for solving a quadratic equation

We've seen that a quadratic equation may have two, one or no roots. How can we find out where they are? We'll look at three different methods:

1. Using a graph to estimate the roots
2. Factoring
3. The quadratic formula.

1. Using a graph to estimate the roots

If we draw a graph of a quadratic function, we can **estimate** the roots by simply looking to see where the parabola crosses the x-axis. For example, the graph of equation 3 in figure 3.13 seems to show that the root is at $x = 2$. In this case it's easy to check that this is correct:

$2^2-4x2+4 = 4-8+4 = 0$ ✓

Indeed, $x = 2$ is a root of equation 3.

In most cases the roots of a quadratic equation won't be whole numbers like this, so normally we won't be able to get the *exact* value from the graph. However, if we make the graph quite large and draw it carefully, we can get quite a good estimate of the root.

Activity 3.10

Estimating roots

Consider the quadratic equation:

$x^2+2x-2 = 0$ (equation 5)

Plot a graph of the left-hand side of this equation, with x going from -3 to 3. Make your diagram quite large: use at least half a sheet of paper. See where the parabola crosses the x-axis, and hence estimate the roots of this equation.

Factoring

Sometimes we might be able to simplify a quadratic function by writing it as two brackets multiplied together, rather as we did in section 1.8.

For example, consider *equation 2: $x^2-2x-3 = 0$*.

We can rewrite the left-hand side as:

$x^2-2x -3 = (x+1)(x-3)$.

To check this, we can multiply out the bracket version:

$(x+1)×(x-3) = x+x+1×x+x×(-3)+1×(-3) = x^2+x-3x-3 = x^2-2x-3$

So, we can rewrite the original *equation 2* as:

$(x+1)(x-3) = 0$.

The brackets are called **factors** of the original quadratic function.

Now, if we multiply together two things and get 0, the *only* way that this can happen is if one of the things is 0. So, in this case, if we know that:

$(x+1)(x-3) = 0$

then we *must* have either:

$(x+1) = 0$ in which case $x = -1$

or $(x-3) = 0$ in which case $x = 3$

Thus, $x = 3$ and $x = -1$ are both roots of the original *equation 2*, as we saw before.

Unfortunately, there's no easy way to spot how to rewrite a quadratic equation using factors like this. However, if we *are* able to do it, it provides an easy way to find the roots.

Activity 3.11

Factorising a quadratic

Solve each of the following quadratic equations by rewriting the left hand side as two factors:

1. $2x^2 + 3x - 2 = 0$ (Hint: multiply out $(x + 2)(2x - 1)$: is this useful?)
2. $x^2 - 4x + 3 = 0$ (Hint: multiply out $(x - 3)(x - 1)$)
3. $6x^2 - x - 1$ (Hint: one of the factors is $(2x - 1)$).

3.7 Square roots

Given a number x, suppose we can find another number s such that $s^2 = x$. Then s is called the **square root** of x.

We write $\sqrt{x}$ to mean the square root of x, so $s = \sqrt{x}$.

For example, $3^2 = 9$ so 3 is the square root of 9. We can write this as $3 = \sqrt{9}$.

You can find square roots on a calculator using the 'square root' button, which will probably be labelled '$\sqrt{x}$'.

Different calculators use this button in different ways. For example, to calculate $\sqrt{9}$ you probably need to press either:

$$\boxed{9} \quad \boxed{\sqrt{x}} \qquad \text{or} \qquad \boxed{\sqrt{x}} \quad \boxed{9} \quad \boxed{=}$$

to get 3. Read the manual for your calculator, or just try both methods, to see which one you should use.

On some calculators you may need to press a button marked 'Shift' or 'Inv' or '2nd' to use the square root function. For example, using the Microsoft Windows Calculator, in its scientific view, you need to click:

$$\boxed{9} \quad \boxed{\text{Inv}} \quad \boxed{x\char`^2} \quad \boxed{=}$$

to find the square root of 9.

Activity 3.12

Square roots

Use your calculator to find the following:

1. $\sqrt{4}$
2. $\sqrt{2}$
3. $\sqrt{20}$
4. $\sqrt{200}$
5. $\sqrt{3.5}$

We've seen that if we square a negative number, we get a positive number. This means that any positive number has *two* square roots. If s is a square root of x, so is $-s$.

For example, we know that 3 is a square root of 9. This means that -3 must be a square root as well:

$$(-3)^2 = (-3) \times (-3) = 9 \checkmark$$

Since calculators have to give a single answer, they're designed to give just the positive square root. This is why in the above examples the calculator displays '3' rather than '-3'. However, it's important to remember that there's a negative square root as well. As a reminder of this, we sometimes write $\pm\sqrt{x}$ to mean 'either the positive square root or the negative square root'.

We've seen that the square of a positive number is positive, and the square of a negative number is also positive. This means that a negative number *doesn't have* a square root. For example, there is *no* number s such that $s^2 = -9$, so -9 doesn't have a square root.

Activity 3.13

Square roots 2

What happens if you try to calculate $\sqrt{-9}$ on your calculator?

Note that 0 is a special case. $0^2 = 0$, so 0 is its own square root, but it does not have any other square root.

3.8 The quadratic formula

If we want to find the roots of a quadratic equation exactly (rather than simply estimating them) but we can't see how to rewrite it using factors, then we can use the **quadratic formula**.

Suppose we have a quadratic equation in standard form:

$$ax^2 + bx + c = 0$$

First we need to calculate the value of:

$$b^2 - 4ac$$

This is called the **discriminant** of the quadratic equation.

One of three things can happen: the discriminant can be negative, 0 or positive.

If the discriminant is negative

If the discriminant is negative, the quadratic equation has no roots.

For example, *equation 4* is:

$$x^2 - 2x + 3 = 0$$

Here $a = 1$, $b = -2$ and $c = 3$. The discriminant is:

$$b^2 - 4ac = (-2)^2 - 4 \times 1 \times 3 = 4 - 12 = -8$$

This is negative, so the equation has no roots, as we saw before.

Activity 3.14

Negative discriminants

Show that each of the following quadratic equations has no roots:

1. $3x^2 + 2x + 1 = 0$

2. $x^2 - 3x + 3 = 0$

3. $x^2 = -1$

If the discriminant is 0

If the discriminant is 0, the quadratic equation has a single root (sometimes called a **repeated root**). We can calculate the value of the root directly. It is:

$$\frac{-b}{2a}$$

For example, *equation 3* was:

$$x^2 - 4x + 4 = 0$$

Here $a = 1$, $b = -4$ and $c = 4$. The discriminant is:

$$b^2 - 4ac = (-4)^2 - 4 \times 1 \times 4 = 16 - 16 = 0$$

So the equation has a single root. This root is:

$$\frac{-b}{2a} = \frac{-(-4)}{2 \times 1} = \frac{4}{2} = 2$$

as we saw when we looked at estimation.

Warning: be careful to get the sign of -b right!

Activity 3.15

Zero discriminants

Show that each of the following quadratic equations has a single root, and find this root:

1. $x^2 + 2x + 1 = 0$

2. $x^2 - 4x + 4 = 0$

3. $x^2 + 6x = -9$.

If the discriminant is positive

If the discriminant $b^2 - 4ac$ is positive, it has two square roots – a positive one and a negative one. We can write these as:

$\pm\sqrt{b^2 - 4ac}$ (recall that '±' means either + or -).

In this case the quadratic equation has two roots. Again we can calculate these roots directly, using the formula:

$$\frac{-b \pm \sqrt{b^2 - 4ac}}{2a} \qquad \text{so the roots are } \frac{-b - \sqrt{b^2 - 4ac}}{2a} \text{ and } \frac{-b + \sqrt{b^2 - 4ac}}{2a}$$

This is called the **quadratic formula**.

It's best to work it out in two stages: find the square roots of the discriminant first, and then put them into the formula to find the two roots of the original equation.

For example, *equation 2* is x^2-2x-3 = 0. Here a = 1, b = -2 and c = -3. The discriminant is:

$$b^2\text{-}4ac = (\text{-}2)^2\text{-}4\times1\times(\text{-}3) = 4\text{-}(\text{-}12) = 4+12 = 16$$

This is positive, so it has two square roots. They are 4 and -4. So one root of the original equation is:

$$\frac{-b+\sqrt{b^2-4ac}}{2a} = \frac{-(-2)+4}{2\times1} = \frac{2+4}{2} = \frac{6}{2} = 3$$

and the other root is:

$$\frac{-b-\sqrt{b^2-4ac}}{2a} = \frac{-(-2)-4}{2\times1} = \frac{2-4}{2} = \frac{-2}{2} = -1$$

So the roots are x = 3 and x = -1, as we saw before.

Using a calculator for the quadratic formula

The quadratic formula $\dfrac{-b\pm\sqrt{b^2-4ac}}{2a}$ is a fraction with two parts.

The top part (the **numerator**) is: $-b\pm\sqrt{b^2-4ac}$

and the bottom part (the **denominator**) is: $2a$

To get the correct value we need to work out the whole of each part, *then* divide one by the other. It's particularly important to remember this when using a calculator.

One way to keep things clear is to write out the calculation in several stages. First we put the values of b and the discriminant into the top part, and the value of a into the bottom part. Then we work out the top part and the bottom part. Then we divide one by the other.

For example, in the last calculation in the previous section we did this as follows:

$$\frac{-b-\sqrt{b^2-4ac}}{2a} \quad \overset{\text{substitute}}{\underset{\text{values}}{\longrightarrow}} \quad \frac{-(-2)-4}{2\times1} \quad \overset{\text{work out top part}}{\underset{\text{and bottom part}}{\longrightarrow}} \quad \frac{-2}{2}$$

$$\overset{\text{divide top part}}{\underset{\text{by bottom part}}{\longrightarrow}} \quad \text{-1}$$

You can also do this calculation in one step on a calculator. However, you must be careful to make the calculator work out the top and bottom parts before doing the division. The way to do this is to include extra brackets like this:

$$\frac{\left(-b-\sqrt{b^2-4ac}\right)}{(2a)} \quad \text{so, for example,} \quad \frac{\left(-(-2)-4\right)}{(2\times1)}$$

when typing the formula into the calculator.

Experiment with your calculator to check that you can use this method to work out the quadratic formula. If you have problems, you can always do the calculation in stages instead.

Example 3.4

Consider *equation 2*: $x^2-2x-3=0$

We saw that the roots are 3 and -1, using the quadratic formula. Check that you can do this calculation on your calculator.

Activity 3.16

Quadratic formula 1

Use the quadratic formula to solve *equation 5:*

$x^2+2x-2=0$

Compare your answer with the estimate that you made in activity 3.6.

Activity 3.17

Quadratic formula 2

Use the quadratic formula to solve the following quadratic equations:

1. $2x^2+5x+2=0$

2. $x^2=x+1$

3.9 Cubic functions and other polynomials

We've seen that a linear function is one involving simply terms like x, and a quadratic function is one involving terms like x^2. A *cubic function* is one that also involves $x^3 = x \times x \times x$. In general a cubic function is one of the form:

$y = ax^3+bx^2+cx+d$

As before we can draw a graph of a cubic function by picking a range of values of x, working out the corresponding y values, plotting the points, and then joining them up to form a smooth curve. We can get various possible shapes:

Figure 3.15: Cubic functions

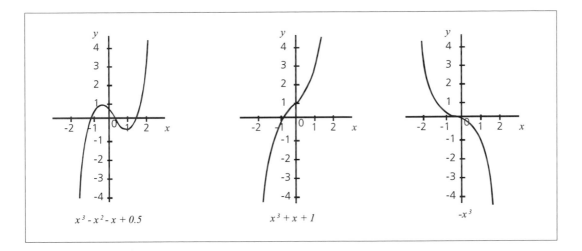

We've seen that a quadratic function either goes up at both ends (i.e. for large negative x at the left-hand end and large positive x at the right-hand end) or it goes down at both ends. A cubic function, on the other hand, always goes up at one end and down at the other. The sign of the x^3 term determines which side goes up: if the x^3 term is positive, then the curve goes up at the right (as in the first two graphs in figure 3.15), and if the x^3 term is negative then the curve goes down at the right (as in the third graph in figure 3.15).

In between a cubic function may reverse direction for a section, as in the first graph in figure 3.15 which goes up then down then up. In other cases it always moves in the same direction, as in the second graph in figure 3.15 which goes up all the time.

We've seen that a quadratic equation (an equation involving a quadratic function) may have 0, 1, 2 roots (values of x where the function is 0). A cubic equation may have 1, 2 or 3 roots, but it must have at least 1 and it cannot have more than 3. The reason it must have at least one root is that it goes up at one end and down at the other, so it must cross the axis in between.

Example 3.5

The cubic equation:

$$x^3 - 7x^2 + 14x - 8 = 0$$

has three roots: 1, 2 and 4.

Check:

$$1^3 - 7 \times 1^2 + 14 \times 1 - 8 = 1 - 7 + 14 - 8 = 0$$

$$2^3 - 7 \times 2^2 + 14 \times 2 - 8 = 8 - 28 + 28 - 8 = 0$$

$$4^3 - 7 \times 4^2 + 14 \times 4 - 8 = 64 - 112 + 56 - 8 = 0$$

The cubic equation:

$$x^3 + x^2 - x - 1 = 0$$

has two roots: −1 and 1.

Check:

$$1^3 + 1^2 - 1 - 1 = 1 + 1 - 1 - 1 = 0$$

$$(-1)^3 + (-1)^2 - (-1) - 1 = -1 + 1 + 1 - 1 = 0$$

The cubic equation:

$$x^3 = 8$$

has only one root: 2.

As with a quadratic function, if we're able to write a cubic function as a product of factors then we can easily see what the roots are.

Consider the cubic equation:

$$2x^3 + 5x^2 - 4x - 3 = 0.$$

We can rewrite the left-hand side as $(x - 1)(x + 3)(2x + 1)$. Check:

$$(x - 1)(x + 3)(2x + 1)$$

$$= (x^2 - x + 3x - 3)(2x + 1) = (x^2 + 2x - 3)(2x + 1)$$

$$= 2x^3 + 4x^2 - 6x + x^2 + 2x - 3 = 2x^3 + 5x^2 - 4x - 3$$

So if this is 0 then the first bracket must be 0, in which case $x = 1$; or the second bracket must be 0, in which case $x = -3$; or the third bracket must be 0, in which case $x = -\frac{1}{2}$.

So this cubic equation has three roots:

−3, -½ and 1

This isn't really a practical method for solving cubic equations, since it's usually very hard to spot what the factors should be. There is a method something like the quadratic formula for cubic equations, but it's much too complicated to cover here.

We can go on adding terms involving x^4, x^5 and so on. In general a function like this – where each term is a constant times a (positive) power of x – is called a **polynomial** in x. The highest power of x is called the **degree** of the polynomial. For example, this is a polynomial of degree 5:

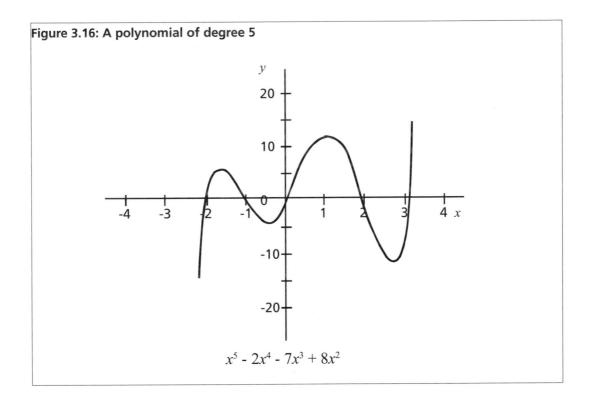

Figure 3.16: A polynomial of degree 5

$$x^5 - 2x^4 - 7x^3 + 8x^2$$

If the degree is **even**, then the polynomial goes up at both ends or down at both ends, like a quadratic function.

If the degree is **odd**, then the polynomial goes up at one end and down at the other, like a cubic function.

If the degree n is even, then the polynomial has between 0 and n roots. If the degree n is odd then the polynomial has between 1 and n roots. Finding the roots exactly is normally difficult or impossible (unless we happen to be able to find factors of the polynomial). However, there are various numerical methods to find the roots approximately.

Polynomials such as cubic functions are often used to approximate other functions, in computer graphics applications for example. If we want to represent a curve, then we can do this by approximating sections of the curve using polynomials. The final curve won't be *exactly* the same as the original one, but it can be very close. The advantage is that to store a polynomial we just need to store its coefficients a, b, etc rather than the whole curve, which is very efficient.

3.10 Logarithmic and exponential functions

In chapter 1 we saw how to extend the idea of an index (such as the 2 in x^2) to cases where the index is negative or fractional. We can use this idea to define a function where the variable appears in the index, so for example:

$$y = 2^x$$

So if $x = 2$ then $y = 4$, if $x = 3$ then $y = 8$, and so on. Furthermore, using the ideas from chapter 1:

if $x = -2$ then $y = \dfrac{1}{4}$, if $x = 0$ then $y = 1$, if $x = \dfrac{1}{2}$ then $y = \sqrt{2}$ and so on.

So, as usual we can pick a range of values of x, work out the corresponding y values, plot the points, and then join them up to form a smooth curve:

Figure 3.17: The function $y = 2^x$

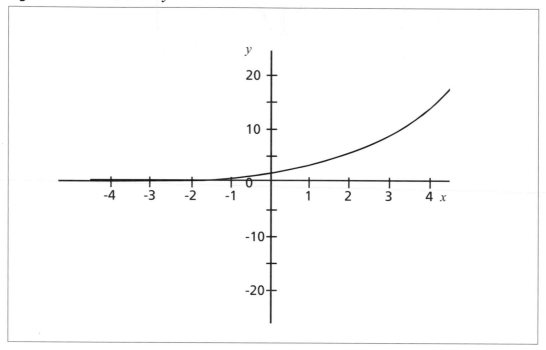

This is an example of an **exponential function**, which means a function of the form $y = a^x$, where a is a constant (2 in the above case). Exponential functions are useful because they model situations in which a quantity's growth is proportional to the value of the quantity. For example, if we have a population of rabbits then, roughly speaking, the birth rate is proportional to the population. This means that the population will follow an exponential function (until the rabbits run out of grass). As you can see from figure 3.17 an exponential function grows rather slowly at first (for negative x) but then starts to grow very quickly for positive x, and soon shoots off the top of the graph.

A particularly important value of the constant a is a number known as e or the **exponential constant**. Its value is approximately

$e = 2.718281828\ldots$

(It looks as though the pattern 1828 keeps repeating here, but in fact it doesn't: the next digit is something different.)

The number e is important because it crops up in lots of areas of mathematics. For example:

- Suppose that the students in a large class each hand in an assignment, but after marking them the tutor hands them back at random, one per student. The chance that *noone* gets their own assignment back is $\dfrac{1}{e}$ or about 0.37

- The curve made by a rope hanging between two points, or by a ship's sail blown by the wind, is called a **catenary**. Its formula is $A(e^{Bx}+e^{-Bx})$ (so there are two exponential functions here)
- Lots of problems to do with growth rates involve e, for example the rate of spread of a new technology or a disease, interest rates, radioactive decay.

Because the number e is so important, the function e^x is sometimes called *the* exponential function:

Figure 3.18 The function $y = e^x$

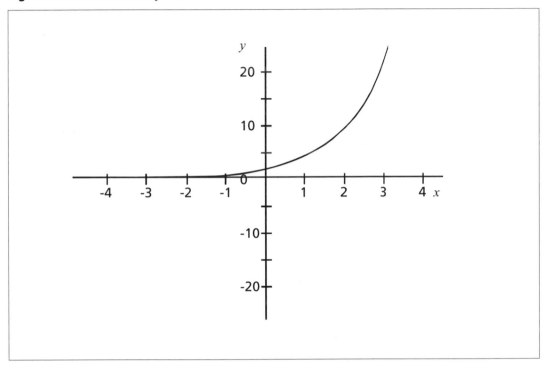

As you can see, e^x looks very similar to 2^x.

Activity 3.18

Exponential functions

Calculate the following using a scientific calculator (there might be one provided as part of your computer's desktop):

1. $2^{1.5}$

2. $3^{2.5}$

3. $e^{1.5}$

4. $e^{-1.5}$

We can define a function which is the opposite or **inverse** of the exponential function. This function is called the **logarithm** function, abbreviated to log. So for example we have:

$$y = 2^x \quad \leftrightarrow \quad x = \log_2 y$$

The subscript (2 in this case) shows which exponential function we're using. In fact the most common values are 10 (because we count in base 10) and e (because it's a useful number). Logarithms using e are called **natural logarithms**, and often abbreviated to ln rather than $\log_e$. So we have:

$$y = 10^x \quad \leftrightarrow \quad x = \log_{10} y$$

and

$$y = e^x \quad \leftrightarrow \quad x = \ln y$$

Scientific calculators often have buttons for both these two logarithm functions.

As usual we can plot a graph of the logarithm function by plotting some points and joining them up with a smooth curve:

Figure 3.19: The function $y = \ln x$

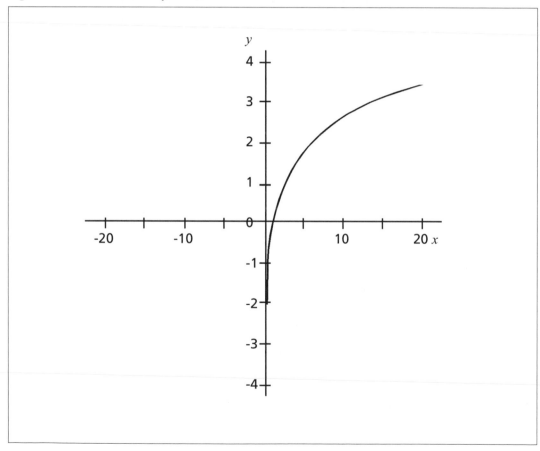

Notice how if we take the graph for $y = e^x$ (figure 3.18) and reflect it in the line $y = x$ then we get the graph for $y = \ln x$ (figure 3.19). This works because the functions are inverses of each other. Notice also that we can't draw ln x for negative x, because e^x is never negative.

Activity 3.19

Logarithms

Calculate the following using a scientific calculator (or computer):

1. $\ln(2)$
2. $\log_{10}(2)$
3. $\log_{10}(10)$
4. $\ln(23)$.

If we take the rules for indices for chapter 1, then we can convert them into rules for logarithms. For example, suppose $a = e^c$ and $b = e^d$. Then $c = \ln a$ and $d = \ln b$. Also, by using the laws of indices we have:

$$ab = e^c e^d = e^{c+d}$$

But this just says that the logarithm of ab is $c + d$. Then, since $c + d = \ln a + \ln b$ we have:

$$\ln(ab) = \ln a + \ln b$$

Indeed, one of the original reasons that logarithms were developed, in the days before calculators, was because this law means they can be used to help do multiplication.

Using the other index laws in a similar way it can be shown that:

$$\ln(a \div b) = \ln a - \ln b$$

$$\ln(a^b) = b\ln a$$

3.11 Summary

In this chapter we looked at more complicated quadratic equations, and saw how to use a graph to see how they behaved. We saw how to decide how many solutions a quadratic equation has, and some ways to find these solutions. We looked at polynomial, exponential and logarithmic functions, using graphs. Quadratic and other equations are useful for studying many different real-world problems.

3.12 Review questions

Question 3.1: For each of the following equations: rearrange the equation into the form $y = mx + c$; predict whether the line will slope upwards or downwards and where it will cross the y-axis; pick two values of x and work out the corresponding values of y; draw sensible axes and plot the line.

1. $x + y = 4$
2. $\dfrac{x}{3} + 2y = 2$

Question 3.2: Finding the roots of a quadratic function 1

Consider the quadratic function $y = 2x^2-3x-4$.

1. Work out the value of y for various small values of x, say $x = -2,-1,0,1,2$.

2. Based on the values that you calculated in part 1, can you tell how many roots this equation has?

   Can you use the values to get a rough idea of where the roots of the curve are? Are they below the x values you used? Or above? Or something else?

3. Using your conclusions from part 2, choose a sensible range of values for x, and plot a graph of the function for this range of values. Use at least half a sheet of paper for your axes.

   You should use the y values that you found in part 1, but you will probably need to calculate y values for some other values of x as well to get a set of points throughout your range. Join up the points with a smooth curve to see the graph.

4. Use the graph to estimate the position of the roots.

5. Use the quadratic formula to calculate the position of the roots. How good were your estimates in part 4?

Question 3.3: Finding the roots of a quadratic function 2

Repeat activity 3.2 for the function:

$$y = 2x+x^2+5$$

as far as possible.

Question 3.4: Factoring

Consider the equation $x^2-4x = 0$.

1. Rewrite the left-hand side of this equation as a product of two factors (i.e. as two things multiplied together).

   Hint: both terms of the left-hand side include an x, so x is a factor.

2. Use part 1 (above) to find the roots of the equation.

Question 3.5: The path of a ball

Lee stands at a point on the ground (the x-axis) somewhere near $x = -2$ and throws a ball. The ball's path is given by the quadratic function $y = 15+3x-x^2$.

1. How high is the ball when it passes the point $x = 1$?

2. Where is Lee standing? Where does the ball land?

   Hint: use the quadratic formula.

3. How far away from Lee does the ball land?

Review question 3.6:   Evaluate the following:

1. $x^3 + x^2 - 2x + 3$ when $x = 2$
2. $x^6 + 7x^3 - 2x$ when $x = -1$
3. x^{10} when $x = 1.5$

 Review question 3.7:   Check the following using a scientific calculator (or computer):

1. $\ln(e^3) = 3$
2. $\ln(6) = \ln(2) + \ln(3)$
3. $\ln(8) = 3 \times \ln(2)$

3.13 Answers to review questions

Question 3.1

1. $x + y = 4 \rightarrow y = x - 4$. The gradient is 1 so the line will slope upwards fairly steeply. The intercept is –4 so the line will cross the y-axis at $y = -4$.

Two sensible values of x might be $x = 0 \rightarrow y = -4$ and $x = 4 \rightarrow y = 0$.

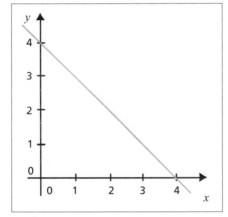

2. $\dfrac{x}{3} + 2y = 2 \rightarrow y = -\dfrac{x}{6} + 1$.

The gradient is $-\dfrac{1}{6}$ so the line will slope upwards very gently. The intercept is 1 so the line will cross the y-axis at $y = 1$.

Two sensible values of x might be $x = 0 \rightarrow y = 1$ and $x = 10 \rightarrow y = -\dfrac{2}{3}$.

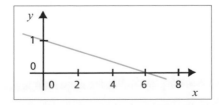

Question 3.2: Finding the roots of a quadratic function 1

1. The y values are:

x	-2	-1	0	1	2
y	10	1	-4	-5	-2

2. Consider what happens as we move through these x values:

As we move from $x = -2$ to $x = -1$ the y value falls from 10 to 1. The coefficient of x^2 is 2, which is positive, so we know that the parabola is convex. So, the first of these points, where $x = -2$, (and probably the second one where $x = -1$ as well) must be to the left of the centre – we have not yet got to the point where the parabola starts rising again.

As we move from $x = -1$ to $x = 0$ the y value changes from 1, which is positive, to -4, which is negative. Since a quadratic function forms a smooth curve it must therefore cross the x-axis (where $y = 0$) somewhere between $x = -1$ and $x = 0$. This means that there must be a root in this region.

We've found one root, but since there are y values both above and below the x-axis, we know that the parabola does *not* just touch the x-axis, so this cannot be a single root – there must be another root somewhere.

As we move from $x = 0$ to $x = 1$ the y value does not change much, so we must be near the centre of the parabola. In fact, since the y value continues to fall during this step, the centre must be somewhere near $x = 1$.

As we move to $x = 2$ the y value is still negative, so we have not yet reached the second root. On the other hand, when we move from $x = 1$ to $x = 2$ the y value goes *up* from -5 to -2. We must, therefore, be beyond the centre of the parabola, so hopefully we may be getting close to the second root. So, overall we know that there are two roots. One is between $x = -1$ and $x = 0$, and the other is a bit above $x = 2$. The centre of the parabola is somewhere near $x = 1$.

3. It seems that a sensible range of x values to use is something like $x = -2$ to $x = 4$, although if you used a slightly different range it doesn't matter. We can extend our table of values to include this range, and to get a good plot we should also add in some extra points near the centre and some near where we think the roots are:

x	-2	-1	-0.5	0	0.5	1	1.5	2	2.5	3	4
y	10	1	-2	-4	-5	-5	-4	-2	1	5	16

Again, if you picked a different sets of points it doesn't matter. If we plot these points, then join them up with a smooth curve, we get something like this:

4. From the graph, the roots appear to be at $x = -0.8$ and $x = 2.4$, approximately.

5. The quadratic equation is $2x^2-3x-4 = 0$ so we have $a = 2$, $b = -3$ and $c = -4$. The discriminant is:

$b^2-4ac = (-3)^2-4\times2\times(-4) = 9-(-32) = 41$

This is positive, so the equation has two roots (confirming what we found in part 2). Substituting this value of the discriminant into the quadratic formula, we have:

$$\frac{-b+\sqrt{b^2-4ac}}{2a} = \frac{-(-3)-\sqrt{41}}{2\times2} = \frac{3+6.4031}{4} = 2.351$$

and

$$\frac{-b-\sqrt{b^2-4ac}}{2a} = \frac{-(-3)-\sqrt{41}}{2\times2} = \frac{3-6.4031}{4} = -0.851$$

The two roots of the equation are thus $x = -0.851$ and $x = 2.351$, so our estimates in part 4 were reasonably good.

Question 3.3: Finding the roots of a quadratic function 2

1. It's sensible to rearrange the function into standard form straight away, to get:

   $y = x^2+2x+5$ (equation 6)

   Using the same values of x as in activity 1, the y values are:

x	-2	-1	0	1	2
y	5	4	5	8	13

2. The coefficient of x^2 is 1, which is positive, so we know that the parabola is convex.

   As we move from $x = -2$ to $x = -1$ the y value falls from 5 to 4, so we know that the point where $x = -2$ is to the left of the centre – we have not yet got to the point where the parabola starts rising again. However, as we move from $x = -1$ to $x = 0$ the y value *rises* from 4 to 5. So, we know that the point where $x = 0$ is to the *right* of the centre, since by then the parabola has started rising again. So, we know that the centre is somewhere near $x = -1$. Since the y values in this region are quite large (4, 5, etc), it seems unlikely that the parabola ever drops down far enough to cross the x-axis. It, therefore, seems unlikely that the equation has any roots. To be absolutely sure, we can use the discriminant. From the standard form of the function in equation 6 we see that $a = 1$, $b = 2$ and $c = 5$. So the discriminant is:

   $b^2-4ac = 2^2-4\times1\times5 = 4-20 = -16$.

   This is negative, so the equation has no roots, as we expected.

3. If we plot the graph, we can see that the parabola doesn't cross the x-axis, as we expected:

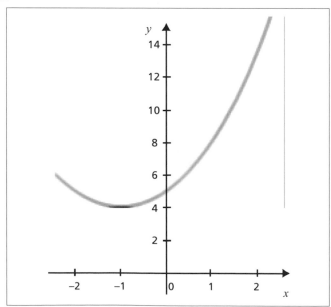

Question 3.4: Factoring

1. The first term x^2 is $x \times x$. The second term $-4x$ is $(-4) \times x$. So both terms contain an x, and we can take out this factor of x as follows:

$$x^2 - 4x = x \times x + (-4) + x = (x + (-4)) + x = (x-4)x$$

2. We can rewrite the original equation as:

$$(x-4)x = 0$$

The only way to multiply two things together and get 0 is for one of them to be 0 to start with. So we must have either:

$(x-4) = 0$, in which case $x = 4$

or, we must have:

$$x = 0$$

Thus, the equation has two roots: $x = 0$ and $x = 4$.

Answer to review question 3.5: The path of a ball

1. At $x = 1$ the height of the ball is:

$$15 + 3 \times 1 - 1^2 = 15 + 3 - 1 = 17$$

2. Again it's sensible to rearrange the equation into standard form to get:

$y = -x^2 + 3x + 15$ (equation 7)

From the standard form of the function in *equation 7* we see that $a = -1$, $b = 3$ and $c = 15$. The discriminant is:

$b^2 - 4ac = 3^2 - 4 \times (-1) \times 15 = 9 - (-60) = 69$

This is positive, so the equation has two roots. These are the point where Lee is standing (because he is on the x-axis) and the point where the ball lands (because it lands on the x-axis). Using the quadratic formula, the roots are:

$$\frac{-b + \sqrt{b^2 - 4ac}}{2a} = \frac{-3 + \sqrt{69}}{2 \times (-1)} = \frac{-3 + 8.3066}{-2} = -2.653$$

and

$$\frac{-b - \sqrt{b^2 - 4ac}}{2a} = \frac{-3 - \sqrt{69}}{2 \times (-1)} = \frac{-3 - 8.3066}{-2} = 5.653$$

We were told that Lee is 'near $x = -2$', so he must be at the root $x = -2.653$.

The ball lands at the other root; at $x = 5.653$.

3. The distance from Lee to where the ball lands is:

$$5.653 - (-2.653) = 8.306$$

Question 3.6

1. $2^3 + 2^2 - 2 \times 2 + 3 = 11$
2. $(-1)^6 + 7 \times (-1)^3 - 2 \times (-1) = 1 + (-7) - (-2) = -4$
3. $(1.5)^{10} = 57.6650\ldots$

Question 3.7 Check the following using a scientific calculator (or computer):

1. $\ln(e^3) = \ln(20.08554\ldots) = 3$
2. $\ln(6) = 1.7918 = 0.6931 + 1.0986 = \ln(2) + \ln(3)$ (with some rounding errors)
3. $\ln(8) = 2.0794 = 3 \times 0.6931 = 3 \times \ln(2)$.

3.14 Feedback on activities

Activity 3.1

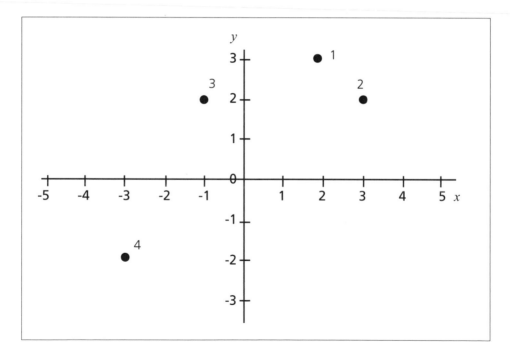

Activity 3.2

To get an accurate plot we should pick an *x*-value at or near each end of the range of *x* we're interested in.

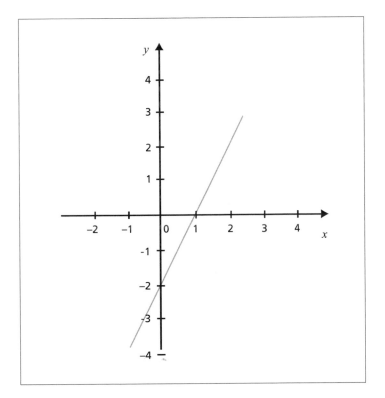

So, if we pick $x = -1$ then the corresponding value of *y* is $2 \times (-1) - 2 = -4$, and if we pick $x = 3$ then the corresponding value of *y* is $2 \times 3 - 2 = 4$.

Thus we know that the points $(-1, -4)$ and $(3, 4)$ lie on the line. So to plot the line we can just plot these two points and join them up.

Activity 3.3

We have the equation:

$y = 3 - 2x.$

The value of m is –2. This is negative, so we expect the line to slope downwards. The value of m is less than –1, so the line will be steep.

The value of c is 3. We expect the line to cross the y-axis at the point $y = 3$.

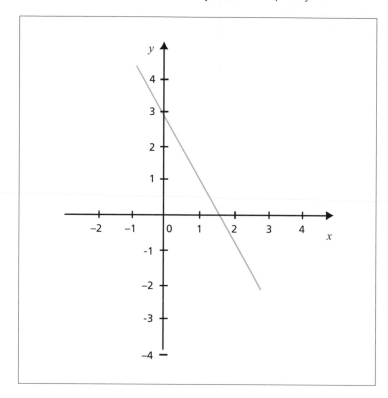

Activity 3.4

Rearrange the following linear equations into the form $y = mx + c$:

1. $10 = 3x + 2y \rightarrow -3x + 10 = 2y \rightarrow y = \dfrac{-3x + 10}{2} = -\dfrac{3}{2}x + 5$

2. $1 = -x - 3y \rightarrow 1 + 3y = -x \rightarrow 3y = -x - 1 \rightarrow y = \dfrac{-x - 1}{3} = -\dfrac{1}{3}x - \dfrac{1}{3}$

3. $-3 = 3y - 4x \rightarrow 4x - 3 = 3y \rightarrow y = \dfrac{4x - 3}{3} = \dfrac{4}{3}x - 1$

Activity 3.5

We have the function

$$10 = 3x + 2y.$$

If $y = 0$ then we have $10 = 3x + 2 \times 0 = 3x$ so $x = \dfrac{10}{3} = 3\dfrac{1}{3}$.

If $x = 0$ then we have $10 = 3 \times 0 + 2y = 2y$ so $y = \dfrac{10}{2} = 5$.

So the line passes through the point $\left(3\dfrac{1}{3}, 0\right)$ (on the x-axis) and the point $(0, 5)$ (on the y-axis).
Plotting these two points and joining them up as usual we get

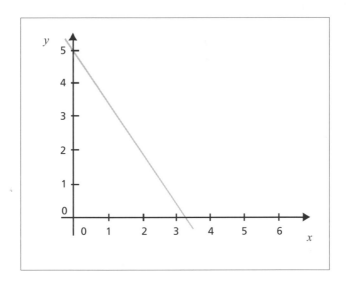

Activity 3.6

x	-3	-2	-1	-0.5	0	0.5	1	2	3
y	9	4	1	0.25	0	0.25	1	4	9

Activity 3.8

1. $2 = 3x + 5x^2 \rightarrow 5x^2 + 3x - 2 = 0$
   So, $a = 5$, $b = 3$ and $c = -2$

2. $x - x^2 + 3x + 1 = 3 + x^2 + 4x - 2 \rightarrow 2x^2 = 0$
   So, $a = 2$, $b = 0$ and $c = 0$

3. $x(x + 2) = 3x - 4 \rightarrow x^2 + 2x = 3x - 4 \rightarrow x^2 - x + 4 = 0$
   So, $a = 1$, $b = -1$ and $c = 4$

4. $-(x + 1)(2 - x) = 3 - x - x^2 \rightarrow -x^2 + 5 = 0$
   So $a = -2$, $b = 0$ and $c = 5$

5. $(1 - x)(4 + x) = (2x - 1)(2x + 2) \rightarrow 3x^2 + 5x - 6 = 0$
   So $a = -3$, $b = 5$ and $c = 6$

Activity 3.9

Substituting $x = -1$ into the equation $x^2-2x-3 = 0$, we get:

$(-1)^2-2\times(-1)-3 = 1-(-2)-3 = 1+2-3 = 0$

However, note that all we've done here is to *check* that certain values were roots, which is easy. The important thing is to be able to *find* these values in the first place. Later on we'll look at some ways to do this.

Activity 3.10

Plotting a graph of the function $y = x^2+2x-2$, we get:

Graph of the function $y = x^2+2x-2$

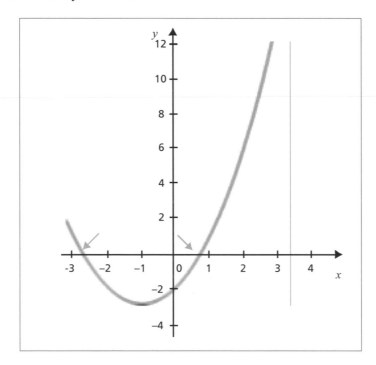

The arrows show where the parabola crosses the x-axis. We can see that the roots are $x = -2.7$ and $x = 0.7$, approximately.

Activity 3.11

1. $(x+2)(2x-1) = 2x^2+4x-x-2 = 2x^2+3x-2$

   so we can rewrite:

   $2x^2+3x-2 = 0$ as $(x+2)(2x-1) = 0$

   For this to be true one of the two factors (brackets) must be 0. So we must have either:

   $(x+2) = 0$, in which case $x = -2$

or

$(2x-1) = 0$, in which case $x = 0.5$.

Thus $x = -2$ and $x = 0.5$ are both roots of this equation. Check:

$2\times(-2)^2+3\times(-2)-2 = 2\times4+(-6)-2 = 8+(-6)-2 = 0$ ✓

and

$2\times(0.5)^2+3\times(0.5)-2 = 2\times0.25+1.5-2 = 0.5+1.5-2 = 0$ ✓

We know that a quadratic equation can have, at most, two roots, so we have found all the roots of this equation.

2. $(x - 3)(x - 1) = x^2 - 3x - x + 3 = x^2 - 4x + 3$ so we can rewrite $x^2 - 4x + 3 = 0$ as $(x - 3)(x - 1) = 0$. Either the first factor $(x - 3)$ must be 0, in which case $x = 3$, or the second factor $(x - 1)$ must be 0 in which case $x = 1$. So the two roots are $x = 1$ and $x = 3$.

3. If one of the factors is $(2x - 1)$, then to get a term $6x^2$ the other factor must be $(3x + a)$, and we need to find a. The x term is $((-1)\times3+2\times a)x = (-3+2a)x$, and this needs to be $-x$ so a must be 1. Check:

$(2x - 1)(3x + 1) = 6x^2 - 3x + 2x - 1 = 6x - x -1$.

So now as usual either the first factor $(2x - 1)$ must be 0, in which case $x = \dfrac{1}{2}$,

or the second factor $(3x + 1)$ must be 0, in which case $x = -\dfrac{1}{3}$.

So the two roots are $x = \dfrac{1}{2}$ and $x = -\dfrac{1}{3}$.

Activity 3.12

These answers are given to six decimal places, but different calculators use different levels of accuracy, and most will show more digits than this:

1. $\sqrt{4} = 2$
2. $\sqrt{2} = 1.414214$
3. $\sqrt{20} = 4.472136$
4. $\sqrt{200} = 14.142136$
5. $\sqrt{3.5} = 1.870829$

Note that: $\sqrt{200} = 10\times\sqrt{2}$.

Can you see why this must be true? *Hint:* square both sides.

Activity 3.13

The calculator should produce some kind of error symbol or error message. This might just be a letter 'E' or the letters 'Err', but the Microsoft Windows XP Calculator, for example, prints the message:

'Invalid input for function.'

Activity 3.14

1. The quadratic equation $3x^2+2x+1 = 0$ is in standard form, so we have $a = 3$, $b = 2$ and $c = 1$. The discriminant is:

   $$b^2\text{-}4ac = 2^2\text{-}4\times3\times1 = 4\text{-}12 = \text{-}8$$

   This is negative, so this equation has no roots.

2. Here $a = 1$, $b = -3$ and $c = 3$, so the discriminant is $(-3)^2 - 4\times1\times3 = -3$ which is negative, so this equation has no roots.

3. Putting the equation into standard form we have $x^2 + 1 = 0$. So here $a = 1$, $b = 0$ and $c = 1$, so the discriminant is $0^2 - 4\times1\times1 = -4$ which is negative, so this equation has no roots.

Activity 3.15

1. The quadratic equation $x^2+2x+1 = 0$ is in standard form, so we have $a = 1$, $b = 2$ and $c = 1$. The discriminant is:

   $$b^2\text{-}4ac = 2^2\text{-}4\times1\times1 = 4\text{-}4 = 0$$

   So this equation has a single root. We can calculate this root:

   $$\frac{-b}{2a} = \frac{-2}{2\times1} = \frac{-2}{2} = -1$$

2. Here $a = 1$, $b = -4$ and $c = 4$, so the discriminant is $(-4)^2 - 4\times1\times4 = 0$, so the equation has a single root.

   The root is $-\dfrac{b}{2a} = -\dfrac{-4}{2} = 2$

3. Putting the equation into standard form we get:

   $$x^2 + 6x + 9 = 0$$

   Here $a = 1$, $b = 6$ and $c = 9$, so the discriminant is $6^2 - 4\times1\times9 = 0$, so the equation has a single root: $-\dfrac{b}{2a} = -\dfrac{6}{2} = -3$

Activity 3.16

The quadratic equation $x^2+2x\text{-}2 = 0$ is in standard form, so we have $a = 1$, $b = 2$ and $c = -2$. The discriminant is:

$$b^2\text{-}4ac = 2^2\text{-}4\times1\times(\text{-}2) = 4\text{-}(\text{-}8) = 12$$

This is positive, so it has two square roots. They are 3.4641 and -3.4641, approximately. Using the quadratic formula we have:

$$\frac{-b+\sqrt{b^2-4ac}}{2a} = \frac{-2+3.4641}{2\times1} = \frac{1.4641}{2} = 0.732$$

or

$$\frac{-b-\sqrt{b^2-4ac}}{2a}=\frac{-2-3.4641}{2\times1}=\frac{-5.4641}{2}=-2.732$$

So the roots are $x = 0.732$ and $x = -2.732$.

Thus, our estimates $x = 0.7$ and $x = -2.7$ in activity 3.10 were quite close, but these calculated values are more exact.

Activity 3.17

1. The equation $2x^2 + 5x + 2 = 0$ is in standard from, so we have $a = 2$, $b = 5$ and $c = 2$. The discriminant is:

$$5^2 - 4\times2\times2 = 25 - 16 = 9.$$

This is positive, so it has two square roots: 3 and -3. Using the quadratic formula we have:

$$\frac{-b+\sqrt{b^2-4ac}}{2a}=\frac{-5+3}{2\times2}=\frac{-2}{4}=-\frac{1}{2}$$

and

$$\frac{-b-\sqrt{b^2-4ac}}{2a}=\frac{-5-3}{2\times1}=\frac{-8}{4}=-2$$

So the roots are $x = -\dfrac{1}{2}$ and $x = -2$.

2. Putting the equation into standard form, we get:

$$x^2 - x - 1 = 0$$

so $a = 1$, $b = -1$ and $c = -1$. The discriminant is:

$$(-1)^2 - 4\times1\times(-1) = 5$$

This is positive, so it has two square roots: 2.2361 and -2.2361, approximately. Using the quadratic formula we have:

$$\frac{-b+\sqrt{b^2-4ac}}{2a}=\frac{-(-1)+2.2361}{2\times1}=1.6180$$

and

$$\frac{-b+\sqrt{b^2-4ac}}{2a}=\frac{-(-1)-2.2361}{2\times1}=-0.6180$$

So the roots are $x = 1.6180$ and $x = -0.6180$, approximately.

Activity 3.18

1. $2^{1.5} = 2.828427\ldots$
2. $3^{2.5} = 15.588457\ldots$
3. $e^{1.5} = 4.481689\ldots$
4. $e^{-1.5} = 0.223130\ldots$

Activity 3.19

1. $\ln(2) = 0.693147\ldots$
2. $\log_{10}(2) = 0.301029\ldots$
3. $\log_{10}(10) = 1$ (of course)
4. $\ln(23) = 3.135494\ldots$

CHAPTER 4

Proportions, percentages and ratios

OVERVIEW

In this chapter we'll look at some basic ideas to do with proportions, percentages and ratios, and they way they measure relative sizes and changes. What does it mean if a system's performance is improved by 10%? What does it mean if inflation is '5%'? How does a bank calculate the interest it pays? How can we scale an image so that the proportions remain correct?

In chapter 1 we looked at basic mathematical operations and simple calculations. In this chapter we'll look at ways of using these techniques to do calculations to help us solve these kinds of problems.

Learning outcomes	On completion of this chapter, you should be able to:

- Convert between numbers, proportions, percentages and ratios

- Calculate simple and compound interest payments

- Convert a nominal interest rate into an APR

- Explain what is meant by π and the golden ratio.

You should have completed chapter 1 before starting this chapter.

4.1 Proportions and percentages

When solving problems, we often deal with changes in quantities, such as the change in a bank balance when interest is paid, the change in the cost of something when inflation occurs or the change in a share price.

For example, suppose that since last week the price of a loaf of bread has risen from 80p to 84p, and the price of a bar of chocolate has risen from 40p to 43p.

A first step is to simply find the change in price for each item: for bread it is 84p-80p = 4p, while for chocolate it is 43p-40p = 3p. The change is larger for the bread, but this is a bit misleading because the overall amount of money involved was larger for the bread, so we might expect larger changes anyway.

Proportions

A more sensible way to compare the changes is to express each change as a **proportion** of the original price, in other words to divide the change by the original price. Thus, we have:

Proportional change for bread: $\frac{4p}{80p} = 0.05$.

Proportional change for chocolate: $\frac{3p}{40p} = 0.075$.

The proportional change is often rather small, as in this case. It can, therefore, be useful to express it as a **percentage**, by multiplying by 100 and writing '%' to indicate that this has been done. So, 0.05 is the same as $(0.05 \times 100)\% = 5\%$ and 0.075 is the same as $(0.075 \times 100)\% = 7.5\%$.

We can convert easily between proportions and percentages:

- If we multiply the proportion by 100, we get the percentage
- If we divide the percentage by 100, we get the proportion.

It's important to note that we will often express numbers such as interest rates as percentages (normally a number between 0 and 100), but when we're using the number in a formula we should express it as a proportion (normally a number between 0 and 1).

Activity 4.1

Proportions and percentages

Express the following changes as proportions, and then as percentages:

1. A change from 40 to 50.

2. A change in price from 87p to 90p.

3. A change in network speed from 11Mb to 12Mb.

4. A change in run-time from 1 minute 40 seconds to 85 seconds.

4.2 Percentage changes

In the previous section we knew the change and expressed this as a percentage. We can also go the other way – if we know the initial value and the percentage change, then we can work out the final value.

For example, suppose a car costs £3,000, but then the price rises by 3%. What is the new cost?

As a proportion, the change is $\frac{3\%}{100} = 0.03$, so as a proportion the new cost is the old cost (1) plus this change (0.03) which would be:

1+0.03 = 1.03. So the new cost is £3,000×1.03 = £3,090.

Now suppose, instead, that the price *falls* by 4% from £3,000.

As a proportion the change is $\frac{-4\%}{100}$ = -0.04, so as a proportion the new cost is the old cost (1) plus this change (-0.04), i.e. 1-0.04 = 0.96. So the new cost is £3,000×0.96 = £2,880.

Activity 4.2

Percentage changes

1. Last year a cake cost £1.25. The inflation rate is 5% per year. How much does the cake cost this year?

2. A sign on a shop says 'Sale: 20% off marked price'. A pair of trousers has a marked price of £35. How much do they cost in the sale?

3. A software company charges a fee of 3% for sending an order. If I order a CD costing £11.99, what is the total cost of the order?

4.3 Ratios

Another way to express a relationship between two quantities is as a **ratio**. Ratios are very closely related to fractions (from chapter 1). If we have two quantities *a* and *b*, then the ratio *a:b* is the value of the fraction:

$$\frac{a}{b}$$

Example 4.1

A computer screen is 32cm wide and 24cm high. The width:height ratio of the screen is:

$$32:24 = \frac{32}{24} = \frac{4}{3}$$

In fact many television screens have a standard width:height ratio of 4:3. The point is that if a picture has width:height ratio 4:3, and our screen has the same width:height ratio 4:3, then we can make the picture larger or smaller and it will exactly fit the screen. If the width is correct then the height will be correct too. This won't be the case if the ratio for the picture and screen are different.

This is why you sometimes see black areas above and below the picture when a film is shown on television: cinema screens and hence cinema films have a different width:height ratio, so it's not possible to fit the picture exactly onto the screen without distorting it. Widescreen televisions often have a width:height ratio of 16:9 instead, since this is a standard ratio for cinemas.

The width:height ratio of a picture or screen is often called the **aspect ratio**.

Activity 4.3

Aspect ratio 1

A widescreen television has a screen 48cm wide, and an aspect ratio of 16:9. How high is the screen?

Activity 4.4

Aspect ratio 2

A laptop screen is 25cm wide and 20cm high. A photograph is 15cm wide and 12cm high. Can the photograph be scaled so that it exactly fits the screen?

Activity 4.5

Aspect ratio 3

Writing paper in Europe is often made in a size called A4. This has the property that if you cut a sheet in half (to produce the size called A5) each half has the same aspect ratio as the original piece. Thus A5 is the same shape as A4. What is the aspect ratio of A4?

Although ratios may appear to be the same thing as fractions, they're useful partly because they can be easier to write. It's easier to write or type 'width:height' than:

$$\frac{\text{width}}{\text{height}},$$ for example.

When using ratios we sometimes use a double colon '::' to mean 'is the same ratio as', so for example:

32:24::4:3 means 32:24 is the same ratio as 4:3, which means $\dfrac{32}{24} = \dfrac{4}{3}$.

Some other areas where ratios are used:

- Schools and universities use the student:staff ratio to indicate the number of students per staff member (on average). Similarly, a company's technical support unit might use the client:technician ratio to indicate the number of clients supported by each technician (on average)
- The important thing about a mixture such as concrete is the ratio of the various ingredients (cement, sand, water)
- A computer system designer might measure the performance of part of the system, such as the fileserver, by calculating the ratio of the time it's working to the time it's idling.

In all these cases the point is that we don't really care how large the numbers are (Is it a big school or a small one? How much cement have we made? How long has the computer system been running?). The important thing is the relative sizes of the two quantities (Have we got the right number of teachers for our students? Is the cement too runny? Is the system working efficiently?).

4.4 **Significant ratios**

π

The ratio of a circle's circumference C to its diameter D is a constant: the ratio doesn't depend on the size of the circle. This ratio is known as π, pronounced *pi*, which is a Greek letter *p* (standing for 'perimeter').

Figure 4.1 The ratio π

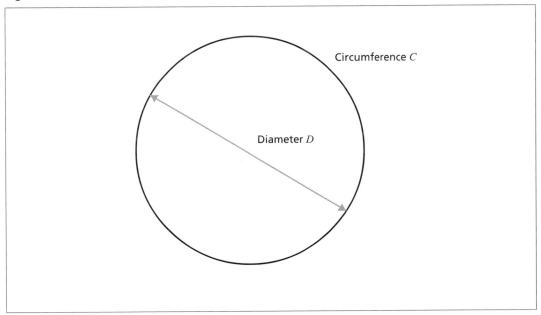

Its value is approximately:

$\pi = 3.14159265358979323846264338327950...$

(the decimal places go on for ever without repeating). π is useful in geometry, but also appears in some probability calculations and various other places.

Activity 4.6

π on a calculator

Most scientific calculators know the value of π, so you don't have to remember it. Does yours have a π button?

The golden ratio

Consider three points on a line:

Figure 4.2: Three points on a line

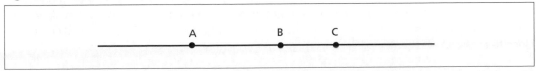

Suppose we want the ratio AB:BC to be the same as the ratio AC:AB (where AB means the distance from A to B, and so on). Where should we put B? The answer to this problem is known as the *golden ratio*. It's sometimes written as φ, pronounced *phi*, which is a Greek letter *f*, or as τ, pronounced *tau*, which is a Greek letter *t*.

To find the value of φ the actual lengths don't matter, because we're interested only in the ratios. So, suppose that the distance AB is 1 and the distance AC is φ. Then the distance BC is $1-\varphi$ and so:

$$AB:BC = \frac{AB}{BC} = \frac{1}{1-\varphi} \quad \text{and} \quad AC:AB = \frac{AC}{AB} = \frac{\varphi}{1}$$

If these ratios are to be the same then we must have:

$$\frac{1}{1-\varphi} = \frac{\varphi}{1} \text{ so } 1 = \varphi(1-\varphi) \text{ so } \varphi^2 - \varphi - 1 = 0$$

This is a quadratic equation, so we know how to solve it from chapter 3. In fact we already *have* solved it, in activity 3.17. The (positive) root was:

$$\varphi = \frac{1+\sqrt{5}}{2} = 1.6180339887498948482045868343656...$$

(again the decimal places go on for ever without repeating).

Activity 4.7

φ on a calculator

Check this calculation on your calculator (although its readout may not give as many decimal places as this).

Note that some people use φ to mean the above value minus 1, i.e. 0.6180... This is the length of AB in the diagram if we take AC to be 1.

The golden ratio is often thought to be the 'ideal' ratio, in some sense, and is felt to be pleasing to the eye. It's therefore used in art, to decide the layout of paintings, for example.

The golden ratio also occurs in nature, for example in the spirals of sea-shells and pine cones.

One more occurrence: consider the Fibonacci sequence:

0, 1, 1, 2, 3, 5, 8, 13, 21, 34, 55, 89, 144, ...

... in which each term is the sum of the two preceding terms (so 13 + 21 = 34, for example). For each term we can work out the ratio of that term to the preceding one. As we go along the sequence these ratios get closer to φ. For example, 144:89 = 1.617977... is already pretty close to φ.

The square root of 2

We've already met √2 in activity 4.5 as the aspect ratio of European A-series paper. Its value is approximately:

$$\sqrt{2} = 1.4142135623730950488001687242097...$$

It doesn't have a standard name or letter, but it does appear occasionally in mathematics. For example, the ratio of the diagonal of a square to a side of the square is √2.

4.5 Simple interest

If one person borrows money from another for a time, they often agree that the borrower will pay the lender a fee or hire charge, called **interest**. The original amount of money borrowed is called the **principal amount**.

A common method is to divide the time into **periods** (often a month or a year), and agree that after each period the borrower will pay the lender a certain proportion of the principal amount. This proportion is called the **interest rate**, and it is often expressed as a percentage.

Example 4.2

Boris borrows £50 from Lia for a year, and they agree on an interest rate of 10% per year.

So, after a year Boris owes interest of 10% of £50 = 0.1×£50 = £5, so he has to pay back the original £50 plus £5 interest = £55.

In practice, the borrower may keep the money for several time periods, and pay all the interest at the end. If they are paying **simple interest**, then the interest is always based on the principal amount. The total amount they owe at a given stage is called the *accrued* **amount**.

Example 4.3

Boris borrows £50 from Lia for three years, and they agree on an interest rate of 10% per year.

So, after one year Boris owes £5 interest as before, so the accrued amount is £50+£5 = £55.

After two years he owes a further 10% of £50 = £5, so the accrued amount is £55+£5 = £60.

Similarly, after three years he owes a further 10% of £50 = £5, so the accrued amount is £60+£5 = £65.

Thus, at the end of three years, he must pay Lia a total of £65.

It's clearer to set out these calculations required in a table (see table 4.1).

Table 4.1 Calculations in a table format – example 4.3

Year	Amount on which interest is calculated	Interest	Accrued amount
1	£50	10% of £50 = 0.1×£50 = £5	£50+£5 = £55
2	£50	10% of £50 = 0.1×£50 = £5	£55+£5 = £60
3	£50	10% of £50 = 0.1×£50 = £5	£60+£5 = £65

Activity 4.8

Simple interest

Boris borrows £200 from Lia for two years, at simple interest of 8% per year. Fill in the blanks in the following table to find the accrued amount at the end of the second year.

Year	Amount on which interest is calculated	Interest	Accrued amount
1	£200		
2	£200		

A formula for simple interest

In this section, we use the following notation:

P = principal amount

A = accrued amount

i = interest rate

n = number of time periods (usually months or years).

We can use a subscript to indicate a given time period, so for example A_2 represents the accrued amount after two time periods.

Then we can calculate the accrued amount using the formula:

$A_n = P \times [1+(i \times n)]$ (formula 1)

Note that interest rates are often expressed in percentages, but in this formula it's important to represent the interest rate, i, as a *proportion* (usually a number between 0 and 1).

Example 4.4

In the transaction in example 4.3 we have:

$P = 50$

$i = 0.1$

$n = 3$

We want to find A_3, the accrued amount after three years. Using the formula, we have

$A_3 = 50 \times [1+(0.1 \times 3)] = 50 \times [1+0.3] = 50 \times 1.3 = 65$

as before.

Note that, as in chapter 1, it's important to work out the part in brackets first. So, first we work out the inner (round) brackets, then the outer (square) brackets, then the whole expression.

Activity 4.9

Simple interest formula 1

Use the simple interest formula to find the accrued amount after two years in *activity 4.8*.

Activity 4.10

Simple interest formula 2

Boris borrows £350 at a simple interest rate of 15%.

Use the simple interest formula to find the accrued amount after 3 years, and after 10 years.

4.6 Compound interest

Example 4.5

Recall example 4.3, in which Boris borrows £50 from Lia for three years, and they agree on an interest rate of 10% per year.

After one year Boris owes interest of £5, so the accrued amount is £55. However, suppose he doesn't actually pay Lia this interest yet, because he plans to pay all the interest at the end. Then during the second year he has effectively borrowed the £5 interest as well as the original £50, so he has borrowed £55.

Thus, they may agree that at the end of the second year he should pay interest on the *whole* of this amount, so the interest will be 10% of £55 = £5.50. The total accrued amount will, therefore, be £55+£5.50 = £60.50.

Similarly, during the third year he has effectively borrowed £60.50: the original £50, plus a total of £5+£5.50 = £10.50 interest that he has not yet paid. Thus, at the end of the third year he will pay interest on the whole of this amount, so the interest will be 10% of £60.50 = £6.05. The total accrued amount will therefore be £60.50 + £6.05 = £66.55.

At the end of three years he must pay Lia a total of £66.55.

Again it's clearer to set the calculations out in a table (see table 4.2).

Table 4.2: Example 4.5 in a table

Year	Amount on which interest is calculated	Interest	Accrued amount
1	£50	10% of £50 = 0.1×£50 = £5	£50+£5 = £55
2	£55	10% of £55 = 0.1×£55 = £5.50	£55+£5.50 = £60.50
3	£60.50	10% of £60.50 = 0.1×£60.50 = £6.05	£60.50+£6.05 = £66.55

This idea is called **compound interest**, where at each stage the interest is based on the accrued amount (as opposed to simple interest, where the interest was always based on the principal amount). The difference is that, as well as paying interest on the principal amount, the borrower is also paying interest on the interest, and so on.

Activity 4.11

Compound interest

Boris borrows £200 from Lia for two years, at compound interest of 8% per year (compare this with activity 4.8, which involved simple interest). Fill in the blanks in the following table to find the accrued amount at the end of the second year.

Year	Amount on which interest is calculated	Interest	Accrued amount
1	£200		
2			

A formula for compound interest

As with simple interest, we can calculate the accrued amount using a formula. With the same notation as before we have the formula:

$$A_n = P \times (1+i)^n \quad \text{(formula 2)}$$

Example 4.6

In the transaction in example 4.5 we have:

$P = 50$, $i = 0.1$ and $n = 3$

So, using the formula we have:

$A_3 = 50 \times (1+0.1)^3 = 50 \times 1.1^3 = 50 \times 1.331 = 66.55$ as before.

To do this calculation on a calculator you will need to use the 'to the power of' button, which will be probably be labelled x^y, y^x, a^b, b^a or $\wedge$. For example, to calculate 3^4, press:

and you should get 81.

Powers on a calculator

Use your calculator to calculate:

1. 5^6
2. 6^5
3. 1.2^3
4. $(1+0.1)^3$

Compound interest formula 1

Use the compound interest formula to find the accrued amount after two years in activity 4.11.

Compound interest formula 2

Repeat activity 4.10 for compound interest rather than simple interest.

4.7 Nominal and effective interest rates

An interest rate is often expressed as a figure per year, but the borrower and lender may agree to calculate the interest more often. For example, a credit card may have a quoted rate of 24%, but with the interest actually calculated as 2% per month.

The quoted annual rate is called the **nominal annual rate**, and the actual rate of interest paid is called the **effective rate** or **annual percentage rate** or **APR**. Because some of the interest is being applied before the end of the year, the effective rate is greater than the nominal annual rate.

Example 4.7

Boris borrows £50 from Lia for a year, and they agree on an interest rate of 8% per year compounded quarterly.

This means that the interest is paid in four parts, calculated each quarter of a year (three months). Each part will be 8%÷4 = 2%. See table 4.3 for a breakdown.

Table 4.3: Example 4.7 – interest paid quarterly

Quarter	Amount on which interest is calculated	Interest	Accrued amount
1	£50	2% of £50 = 0.02×£50 =£1	£50+£1 = £51
2	£51	2% of £51 = 0.02×£51 = £1.02	£51+£1.02 = £52.02
3	£52.02	2% of £52.02 = 0.02×£52.02 = £1.0404	£52.02+£1.0404 = £53.0604
4	£53.0604	2% of £53.0604 = 0.02×£53.0604 = £1.061208	£53.0604+£1.061208 = £54.12 (2dp)

To find the total interest that Boris paid we could add up the quarterly interest figures in the third column, but a simpler method is just to take the total amount he paid and subtract the original amount he borrowed:

£54.12-£50 = £4.12

So, the effective rate of interest or APR is:

$$\frac{£4.12}{£50} \times 100\% = 8.24\%$$

Recall that the nominal rate here is 8%. Thus, as expected, the APR is a bit higher than the nominal rate.

A formula for APR

If we know the nominal annual rate and the number of compoundings in a year, then we can calculate the APR directly. If:

i = nominal annual rate

n = number of compoundings in one year

then we can calculate the APR using the formula:

$$\text{APR} = \left(1+\frac{i}{n}\right)^n - 1 \quad \text{(formula 3)}$$

Note that n appears twice in this formula.

Example 4.8

In the transaction in example 4.7 we have:

$i = 0.08$

$n = 4$

so the APR is given by

$$\text{APR} = \left(1 + \frac{1.08}{4}\right)^4 - 1 = (1+0.02)^4 - 1 = 1.02^4 - 1$$
$$= 1.0824322 - 1 = 0.0824322 = 8.24\% \ (2\text{dp})$$

expressed as a percentage.

Example 4.9

If a credit card has a nominal annual rate of 24% compounded monthly (i.e. 12 times per year), then:

$i = 0.24$

$n = 12$

so the APR is

$$\text{APR} = \left(1 + \frac{0.24}{12}\right)^{12} - 1 = (1+0.02)^{12} - 1 = 1.02^{12} - 1$$
$$= 1.268242 - 1 = 0.268242 = 26.82\% \ (2\text{dp})$$

expressed as a percentage.

Activity 4.15

APR 1

1. Suppose a credit card has a nominal annual rate of 24% as before, but now it is compounded weekly (i.e. 52 times per year). What is the APR?

2. Suppose this credit card is compounded daily (assume there are 365 days in a year). What is the APR?

Activity 4.16

APR 2

A finance agreement has a nominal annual rate of 10% which is compounded every two months. What is the APR?

4.8 Summary

In this chapter we saw how to use proportions, percentages and ratios to represent changes. We studied some common ratios, and saw how to use these ideas to calculate various types of interest.

4.9 Review questions

 Question 4.1

1. A network manufacturer claims that their new network is '20% faster than the opposition'. If the opposition's network runs at 20mb then how fast is the new network?

2. A computer consultant claims that his system can cut a company's costs by 5%. If its costs are cut from £22m to £20m, is the claim true?

Question 4.2: Percentages

Suppose that the government introduces a toaster tax at a rate of 10%, and the price of a toaster in the shops is £20. If the toaster tax now rises to 12%, but the basic cost of a toaster stays the same, what will the new price in the shops be?

Hint: work in two stages as follows:

1. Before the tax rise, find the basic cost of the toaster, i.e. the cost before the tax is added.

2. After the tax rise, use the basic cost to find the new price of the toaster including tax.

Question 4.3: You're making a cake, and although you can't remember the exact recipe you know that the ratio of flour to butter should be 3:2 and the ratio of flour to fruit should be 3:1. If you have 100g of flour (and want to use all of it), how much butter and fruit should you use?

Question 4.4: If you borrow £20 at a simple interest rate of 6% per year, how much do you owe after:

1. 1 year?

2. 5 years?

3. 100 years?

Question 4.5: Compound interest and APR

Suppose Boris borrows £150 for three years, at a nominal annual rate of 18% compounded monthly. We can work out his accrued debt in (at least) two different ways:

1. We can find the interest rate per month, then say that Boris pays compound interest at this rate per month for 36 months.

2. We can find the APR of this scheme, then say that Boris pays compound interest at this rate for three years.

In each case we can use the *compound interest formula 2* to find the accrued amount of the debt.

Use both these methods to find the accrued amount of the debt. What do you notice?

4.10 Answers to review questions

Question 4.1

1. The percentage change is 20%, so the proportional change is 0.20, so the new rate is 1 + 0.20 = 1.20 of the old rate, so the new rate is 20Mb × 1.20 = 24Mb.

2. The change is £2m, so the proportional change is:

$\dfrac{2}{22} = 0.0909$, so the percentage change is 0.0909 × 100 = 9.09%

This is greater than 5%, so the claim is true.

Question 4.2: Percentages

1. Before the tax rise, with toaster tax of 10%, the price of a toaster is 100%+10% = 110% of the basic cost.

So, expressed as a proportion, the price is:

$\dfrac{110\%}{100\%}$ = 1.1 times the basic cost. The basic cost is therefore:

$\dfrac{£20}{1.1}$ = £18.18

2. After the tax rise, with toaster tax of 12%, the price of a toaster is 100%+12% = 112% of the basic cost. Expressed as a proportion the price is

$\dfrac{112\%}{100\%}$ = 1.12 times the basic cost. The price is therefore £18.18×1.12 = £20.36.

Question 4.3

If *b* is the amount of butter then $\dfrac{100}{b} = \dfrac{3}{2}$ so $\dfrac{b}{100} = \dfrac{2}{3}$ so $b = \dfrac{2 \times 100}{3} = 67g$.

If *f* is the amount of fruit then $\dfrac{100}{f} = \dfrac{3}{1}$ so $\dfrac{f}{100} = \dfrac{1}{3}$ so $f = \dfrac{1 \times 100}{3} = 33g$.

So you should use 67g of butter and 33g of fruit.

Question 4.4

The interest rate is 0.06, so using the simple interest formula 1:

1. $A_1 = 20\times[1 + 0.06\times1] = 20\times1.06 = 21.20$, so you owe £21.20
2. $A_5 = 20\times[1 + 0.06\times5] = 20\times1.3 = 26$, so you owe £26
3. $A_{100} = 20\times[1 + 0.06\times100] = 20\times7 = 140$, so you owe £140

Question 4.5: Compound interest and APR

1. The interest rate per month is:

$$\frac{18\%}{12} = 1.5\% = 0.015$$

The principal amount is P = £150, and the number of time periods (months) is 36. So, using the *compound interest formula 2*, the accrued amount is:

$$A_{36} = 150\times(1+0.015)^{36} = 150\times1.015^{36} = 150\times1.709140 = 256.37$$

2. The nominal annual interest rate is 18% = 0.18 and the number of compoundings in one year is n = 12.   So, using the APR *formula 3*, we have:

$$APR = \left(1+\frac{0.18^{12}}{12}\right)-1 = (1+0.015)^{12}-1 = 1.015^{12}-1$$

$$= 1.195618-1 = 0.195618 = 19.5618\%$$

So, the APR is 19.5618% = 0.195618, the principal amount is P = £150, and the number of time periods (years) is 3. So, using the *compound interest formula 2*, the accrued amount is:

$$A_3 = 150\times(1+0.195618)^3 = 150\times1.1956^3$$

$$= 150\times1.709139 = 256.37$$

Essentially, we have performed the same calculation in two different ways. Both methods are calculating the accrued amount of the debt after three years, so, of course, they give the same value.

4.11 Feedback on activities

Activity 4.1: Proportions and percentages

1. The change is 50-40 = 10.

   As a proportion this is $\frac{10}{40} = 0.25$

   As a percentage this is $\frac{10}{40} \times 100 = 25\%$

2. The change is 90p-87p = 3p.

   As a proportion this is $\frac{3p}{87p} = 0.0345$

   As a percentage this is $\frac{3p}{87p} \times 100 = 3.45\%$

3. The change is 12Mb – 11Mb = 1Mb.

   As a proportion this is $\dfrac{1Mb}{12Mb} = 0.08333$

   As a percentage this is $\dfrac{1Mb}{12Mb} \times 100 = 8.33\%$

4. The change is 1 minute 40 seconds – 85 seconds = 15 seconds.

   We need to express everything in the same unit: either minutes or seconds. It doesn't matter which we pick, although using seconds involves fewer decimal points and awkward calculations, so we'll write 1 minute 40 seconds as 100 seconds.

   As a proportion the change is $\dfrac{-15}{100} = -0.15$

   As a percentage the change is $\dfrac{-15}{100} \times 100 = -15\%$

Activity 4.2: Percentage changes

1. The proportional change is 0.05, so as a proportion the new price is 1+0.05 = 1.05; the new price is £1.25×1.05 = £1.31
2. The proportional change is -0.2, so as a proportion the new price is 1-0.2 = 0.8; the new price is £35×0.8 = £28
3. The proportional change is 0.03, so as a proportion the new cost is 1+0.03 = 1.03; the new cost is £11.99×1.03 = £12.35

Activity 4.3: Aspect ratio 1

If the height of the screen is h then we have

$48{:}h = 16{:}9$ so $\dfrac{48}{h} = \dfrac{16}{9}$ so $\dfrac{h}{48} = \dfrac{9}{16}$ (taking reciprocals) so $h = \dfrac{48 \times 9}{16} = 27$

so the screen is 27cm high.

Activity 4.4: Aspect ratio 2

The aspect ratio of the screen is $25:20 = \dfrac{25}{20} = \dfrac{5}{4}$.

The aspect ratio of the photograph is $15:12 = \dfrac{15}{12} = \dfrac{5}{4}$.

The screen and the photograph have the same aspect ratio, so it *is* possible to scale the photograph so that it exactly fits the screen.

Activity 4.5: Aspect ratio 3

Suppose a piece of A4 paper has short dimension s and long dimension l.

If we cut it in half, we will get a piece of paper with short dimension $\dfrac{l}{2}$ and long dimension s.

The aspect ratios of the two pieces must be the same, so we have:

$$l:s = s:\dfrac{l}{2} \ \text{ so } \ \dfrac{l}{s} = \dfrac{s}{1/2} \ \text{ so } \ \dfrac{l}{s} = \dfrac{2s}{l} \ \text{ so } \ \dfrac{l^2}{s^2} = 2 \ \text{ so } \ \left(\dfrac{l}{s}\right)^2 = 2 \ \text{ so } \ \dfrac{l}{s} = \sqrt{2} \ \text{ so } \ l:s = \sqrt{2}$$

Thus A4 paper has an aspect ratio of $\sqrt{2} = 1.4142...$

As a check: A4 paper is 297mm long and 210mm wide, so its aspect ratio is

$$\dfrac{297}{210} = 1.4142...$$

Activity 4.8: Simple interest

Year	Amount on which interest is calculated	Interest	Accrued amount
1	£200	8% of £200 = 0.08x£200 = £16	£200+£16 = £216
2	£200	8% of £200 = 0.08x£200 = £16	£216+£16 = £232

The accrued amount after two years is £232.

Activity 4.9: Simple interest formula 1

$P = 200, \ i = 0.08, \ n = 2$

So:

$$A_2 = 200\times[1+(0.08\times2)] = 200\times[1+0.16] = 200\times1.16 = 232.$$

Activity 4.10: Simple interest formula 2

$P = 350, i = 0.15$

So:

$A_3 = 350 \times [1+(0.15\times3)] = 350 \times [1+0.45] = 350 \times 1.45 = 507.50$

and

$A_{10} = 350 \times [1+(0.15\times10)] = 350 \times [1+1.5] = 350 \times 2.5 = 875$

Activity 4.11: Compound interest

Year	Amount on which interest is calculated	Interest	Accrued amount
1	£200	8% of £200 = 0.08x£200 = £16	£200+£16 = £216
2	£216	8% of £216 = 0.08x£216 = £17.28	£216+£17.28 = £233.28

The accrued amount after two years is £233.28.

Note: Here we've rounded the amounts to the nearest penny at each stage.

Activity 4.12: Powers on a calculator

1. $5^6 = 15{,}625$.
2. $6^5 = 7{,}776$.
3. $1.2^3 = 1.728$.
4. $(1+0.1)^3 = 1.331$

Activity 4.13: Compound interest formula 1

$P = 200, i = 0.08, n = 2$

So:

$A_2 = 200 \times (1+0.08)^2 = 200 \times 1.08^2 = 200 \times 1.1664 = 233.28$

Activity 4.14: Compound interest formula 2

$P = 350, i = 0.15$

so

$A_3 = 350 \times (1+0.15)^3 = 350 \times 1.15^3 = 350 \times 1.520875 = 532.31$

and

$$A_{10} = 350 \, (1+0.15)^{10} = 350 \times 1.15^{10} = 350 \times 4.04556 = 1,415.95.$$

Feedback on activity 4.15: APR 1

1. $i = 0.24$ and $n = 52$ so:

$$APR = \left(1 + \frac{0.24}{52}\right)^{52} - 1 = (1+0.00461538)^{52} - 1$$

$$= 1.00461538^{52} - 1 = 1.270547 - 1$$

$$= 0.270547 = 27.05\%$$

2. $i = 0.24$ and $n = 365$ so:

$$APR = \left(1 + \frac{0.24}{365}\right)^{365} - 1 = (1+0.000657534)^{365} - 1$$

$$= 1.000657534^{365} - 1 = 1.271149 - 1$$

$$= 0.271149 = 27.11\%$$

Feedback on activity 4.16: APR 2

$i = 0.1$ and $n = 6$ so:

$$APR = \left(1 + \frac{0.1}{6}\right)^{6} = (1+0.0166667)^{6} - 1 = 1.0166667^{6} - 1$$

$$= 1.104260 - 1 = 0.104260 = 10.43\%$$

CHAPTER 5

Sets

OVERVIEW

It is often the case that different objects are categorised in terms of common attributes that they possess. For example, people can be categorised in terms of their gender or perhaps in terms of their age. This is useful to do since certain categories possess the same properties: in the UK, people under the age of 18 cannot vote, for example.

In mathematics we have the notion of a set to deal with this idea. A set is a collection of objects that often (but not necessarily) possess similar properties. In mathematics, sets are the fundamental building blocks upon which everything is founded. In computing, sets are used extensively to gather together objects that have similar properties.

Learning outcomes   On completion of this chapter, you should be able to:

- Describe a set as a collection of elements

- Manipulate sets using unions, intersections and minus

- Decide when one set is a subset of another

- Describe the notion of the Universal set

- Form the complement of a set, given its Universal set.

5.1 Sets – elements

In the overview you found out that a set is a collection of objects or things that usually have something in common. To be a little bit more precise, a set, denoted usually by a capital letter, is a collection of *elements*, listed in no particular order.

Example 5.1

Suppose that C is the set of primary colours, then C consists of the *elements:* red, green and blue.

This is a typical example of a set. The set *C*, in this example, has three elements, red, green and blue. One of the main attractions of using mathematics to describe sets is that we can manipulate the sets themselves to find more information about their elements.

Before we continue, let's first find how to write down a set in terms of its elements properly. It is important, in mathematics, to have standard ways of writing things. This is called *notation*. Notation is like the language of mathematics. In maths we write down a set using curly brackets, {}. So a set is a list of elements inside curly brackets.

In the example above we would write:

C = {red, green, blue}.

Now you have a go.

Activity 5.1

Set basics

Write down the set A that consists of whole numbers from 10 to 16.

Hint: Remember, a whole number is a number without any fractions in it, so 1,2,3, and so on.

As we have said already, the objects in a set are called its elements. To say that something is in a set or is an element of a set, we use the symbol ∈ that looks a little bit like the letter *E*. In example 5.1 we had the set:

C = {red, green, blue}.

So for example you could write:

red ∈ C

to mean that red is an element of the set C.

Activity 5.2

Set elements

Pick an element of the set in activity 5.1 and write down in notation that the element is in the set *A*.

An important property of sets is that elements of a set are never repeated. Consider the next example.

Example 5.2

Let *L* be the set of letters that make up the word 'MISSISSIPPI'.

You might be tempted to write down:

> *L* = {M, I, S, S, I, S, S, I, P, P, I}.

But this would be wrong because you have repeated the letters.

The correct answer is:

> *L* = {M, I, S, P}.

Another important thing to notice is that the elements of a set are written in no particular order; the answer to example 5.2 could have been *L* = {s, i, p, m} and this would have been just as correct.

Activity 5.3

World cup winners

Between 1960 and 1990 the football World Cup was won by the following:

1962	1966	1970	1974	1978	1982	1986	1990
Brazil	England	Brazil	Germany	Argentina	Italy	Argentina	Germany

Write down the set *W* of World Cup winners between 1960 and 1990.

Activity 5.4

Years

Write down the set of years between 2000 and 2005 inclusive.

5.2 Unions

So far you have learned how sets are described in mathematics and how to write down a set in terms of its elements. However, the real usefulness of describing sets in this way comes from the way that you can manipulate them in lots of different ways. The first way that you will learn how to do this is using the union.

If you have two sets, *A* and *B*, then the union of these two sets is another set consisting of all of the elements of both of the sets *A* and *B*. The union of two sets is very like adding the sets together. The important thing to remember though is that the elements of sets are never repeated.

Example 5.3

If *A* is the set of numbers from 1 to 5, and *B* is the set of even numbers from 3 to 9, then:

> *A* = {1,2,3,4,5}
> *B* = {4,6,8}.

So the union of the sets *A* and *B* is the set of all these values:

{1,2,3,4,5,6,8}.

Notice the following points from the example:

1. The elements are only written once; 4 is in both *A* and *B* but we only write it once in the union.

2. The elements are not in any particular order; we could have written {2,6,3,4,8,1,5}.

The union of the two sets is written using the symbol $\cup$ which looks like a *U*. So for the example we had above we could have written:

$$A \cup B = \{1,2,3,4,5,6,8\}.$$

This notation is read as '*A* union *B*'.

Example 5.4

We do not always have similar elements in a set, but the union is still the set of all the values.

Suppose that:

C = {red,green, blue}
X = {12,24,36}.

Then

$$X \cup C = \{red,12,green,24,36,blue\}$$

Activity 5.5

Unions

If

D = {a,b,c,d,e} and

E = {b,d,e,f,g}

what is *D*∪*E*?

Another way of thinking about the union is that the set $A \cup B$ consists of all the elements that are either in *A* or in *B*. This is useful if the sets are described by some property as the next example illustrates.

Example 5.5

If *M* is the set of people who are male and *T* is the set of people who are aged under 20, then the set *M*∪*T* is the set of people who are either male or are under 20.

Notice that *M*∪*T* contains all the males and all the people under 20; in particular, it contains all the males who are under 20.

5.3 Intersections

The union of two sets was a way of creating a new set from two other sets. You saw that the union of a set A and a set B is the set of elements that are either in A or B. Another useful way of creating a set from two other sets is the intersection. If A and B are two sets, then the intersection of A and B is the set consisting of the elements that are common to both A and B, that is, the elements that are in both the sets.

Example 5.6

Suppose that we have the same sets as in example 5.4:

   $A = \{1,2,3,4,5\}$ and
   $B = \{4,6,8\}$.

Then the intersection of A and B is the set

   $\{4\}$

since 4 is the only element in both the sets.

Notation

The intersection is denoted by the symbol $\cap$. So in the example above you would have written:

   $A \cap B = \{4\}$

This is read as 'A intersection B'.

Example 5.7

Suppose that

   $C = \{red,green,blue\}$
   $D = \{red,blue,purple\}$.

then

   $C \cap D = \{red,blue\}$.

The empty set

At this point you must learn about a very special set indeed, called the empty set. The empty set is the set consisting of *no* elements at all. The empty set is given the symbol $\emptyset$, or defined as $\{\}$, and is read as 'the empty set'.

The importance of this set is evident if you go through the following example.

Example 5.8

If

   $C = \{red,yellow,green\}$ and
   $D = \{blue,purple,elephant\}$

then there are no common elements in both of these sets and so

   $C \cap D = \emptyset$.

From the example, you can see that the statement that $A \cap B = \emptyset$ is an important one; it says that there are no elements at all common to both the sets A and B. Two sets that satisfy this are called *disjoint*.

When you learned about the union of two sets, you found out that another way of thinking of the union is the set of elements that are in one, or other or both of the original sets. The intersection of A and B can be thought of as the set of elements that are in A and are also in B.

Example 5.5 (continued)

If M is the set of people who are male and T is the set of people who are aged under 20, then the set $M \cap T$ is the set of males who are under 20.

5.4 Minus

Although minus is something that you will already understand from arithmetic, there is a type of minus that we can define for sets. The set A minus B is the set of all the elements in A that are not in B.

Example 5.9

Let

$A = \{1,2,3,4,5\}$ and

$B = \{4,6,8\}$.

The only element that is in both the sets is 4 and so A minus B is the set:

$\{1,2,3,5\}$.

Notice that we don't care about the other elements in B, only the ones that are common to A.

Notation

We will use the backslash symbol, \, to denote minus. So the set A minus B is written $A \backslash B$ and this is read as 'A minus B' or sometimes 'A without B'. If you look at other textbooks on sets, the minus is sometimes written $A-B$, but we will not write it this way because you should understand that it is not the same as subtraction in arithmetic.

Example 5.10

Suppose that

$C = \{red,green,blue\}$ and

$D = \{red,blue,purple\}$.

Then:

$C \backslash D = \{green\}$.

Try the following activity before going on.

Activity 5.6

Set minus

If

    $A = \{1,2,3,4,5\}$ and

    $B = \{3,4,5,6,7,8\}$.

Then what is $A \backslash B$? What is $B \backslash A$?

Activity 5.7

Set minus 2

Let W be the set of World Cup winners from activity 5.3:

{Brazil, England, Germany, Argentina, Italy}.

Let E be the set

{England, Germany}.

What is $W \backslash E$? What is $E \backslash W$?

Don't forget that if you find this activity difficult then the solutions are at the end of this chapter . If you do need to look at the solutions, then please spend some time getting to grips with the minus as it is important that before you go on you understand this concept.

Important: In the activity above you should find that $A \backslash B$ is not the same set as $B \backslash A$. This is almost always the case. Notice that this is not the case with either union or intersection; for example, $A \cap B$ is the same set as $B \cap A$. We say that the union and intersection are *commutative* because it does not matter which order you do them, whereas minus is said to be *noncommutative*.

5.5 Subsets and supersets

Often when we are using sets we come across sets that are completely contained in others. What this really means is that *every* element of A is also an element of B. When this happens we say that A is contained in B. This gives us a kind of measure of the relative size of sets. If a set A is contained in B, then you should think of it as meaning that A is 'smaller' than B.

Example 5.11

Suppose that

    $C = \{red,yellow,green,blue,purple\}$ and

    $D = \{yellow,blue\}$.

Then D is contained in C since all the elements of D are also elements in C.

It is because of this idea of relative size that if A is contained in B then A is called a **subset** of B (the prefix 'sub' always means that something is lesser or beneath something else), and similarly B is called a *superset* of A.

Actually, you will hardly ever see the word 'superset' used; it is much more common to only use the word 'subset'. In fact, we will not mention the word 'superset' again in this chapter. Why do you think that we only need one definition instead of the two?

Notation

If *A* is a subset of *B*, then we use the symbol $\subset$. So, we could have written in the example that *D*$\subset$*C*. This is read as '*D* is a subset of *C*' or perhaps '*D* is contained in *C*'.

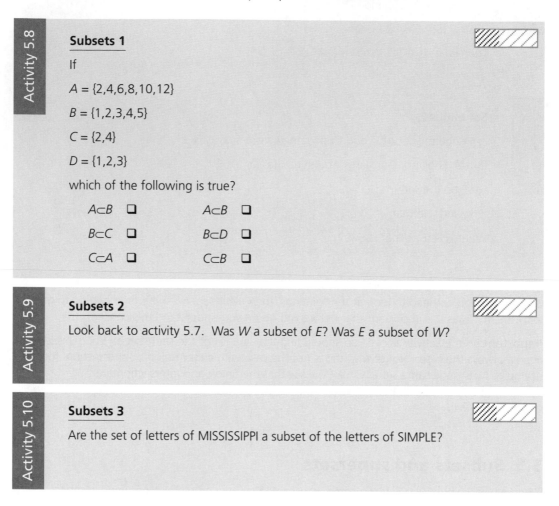

Activity 5.8

Subsets 1

If

A = {2,4,6,8,10,12}

B = {1,2,3,4,5}

C = {2,4}

D = {1,2,3}

which of the following is true?

A$\subset$*B* ❑		*A*$\subset$*B* ❑	
B$\subset$*C* ❑		*B*$\subset$*D* ❑	
C$\subset$*A* ❑		*C*$\subset$*B* ❑	

Activity 5.9

Subsets 2

Look back to activity 5.7. Was *W* a subset of *E*? Was *E* a subset of *W*?

Activity 5.10

Subsets 3

Are the set of letters of MISSISSIPPI a subset of the letters of SIMPLE?

5.6 The Universal set

So far, we've only glanced at the usefulness of sets in both mathematics and in the real world. Sets can be used to give us lots of information about things and are often used in business and computing. For example, sets are one of the principal structures used in computer programming and their manipulation is a very powerful device.

You have found out, so far, how to form the union, intersection and minus of two sets. However, these concepts, by themselves, are a little narrow. In order to broaden our ideas, we need to introduce the concept of a *Universal set*.

A Universal set is a special set that includes all the possible elements you will need in a given situation. So, if you were to form the unions or intersections of two sets, or to find one set minus another, then the resulting sets are all still subsets of the Universal set.

Example 5.12

Suppose that we are using sets that contain letters of the English alphabet as their members, then a Universal set could be the set containing all the letters of the alphabet:

{A,B,C,D,E,F,G,H, I,J,K,L,M,N,O,P,Q,R,S,T,U,V,W,X,Y,Z}.

Notation

The Universal set is always written using a curly 'u', ⊔, and this is simply read as 'the Universal set'.

Example 5.12 (continued)

From example 5.2 we had L = {M, I, S, P}. You could say that:

$L \not\subset U$

Example 5.13

Suppose we are writing a program to simulate the value obtained on throwing die. In this case the Universal set is

⊔ = {1,2,3,4,5,6}.

These are all the possible values.

Special Universal sets

The Universal set is always the biggest set you can think of for the problem you are looking at, and sometimes they can be used to mean that the sets you are looking at contain elements of a certain type.

Example 5.12 (continued): In this example we have:

⊔ = {A,B,C,D,E,F,G,H, I,J,K,L,M,N,O,P,Q,R,S,T,U,V,W,X,Y,Z}

So, if you say that a set $X \subset U$, then you know that the set X contains letters.

There are some other special Universal sets that are useful to know. The first of these is the *natural numbers*. Natural numbers are whole numbers from 1 onwards. All the universal sets in this section have a special letter to write them. For the natural numbers we use the symbol $\mathbb{N}$.

The next special set is the set of *integers*. Integers are, again, whole numbers but this time we include negative whole numbers and 0; we use the symbol $\mathbb{Z}$ for the set of integers.

Next we have the *rational numbers*. Rational numbers are numbers that can be written as fractions; positive or negative. We use the symbol $\mathbb{Q}$ for the rational numbers.

Finally, we have the real numbers. Real numbers are all the possible numbers of which you can think. They include numbers like π or $\sqrt{2}$ which are both numbers that cannot be written as fractions. We use the symbol $\mathbb{R}$ for the real numbers.

Activity 5.11

Special subsets

Explain why the following things are true:

- $\mathbb{N} \subset \mathbb{Z}$

- $\mathbb{Z} \subset \mathbb{Q}$

Hint: Note that for the second one that any whole number, *n*, can be written as the fraction *n*/1.

Note that we have the following inclusions:

$$\mathbb{N} \subset \mathbb{Z} \subset \mathbb{Q} \subset \mathbb{R}$$

5.7 The complement

Once you have decided what the Universal set is, you can define the complement of a set. The complement of a set, *A*, is the set of all possible elements that are *not* in *A*. This set is always useful to be able to find since we can write the statement '*X* is not an element of the set *A*' as '*X* is an element of the complement of *A*'. Why is this true?

Example 5.13

If

$$\mathsf{U} = \{1, 2, 3, 4, 5, 6\} \text{ and}$$
$$D = \{1, 4, 6\}$$

then the complement of *D* is

$$\{2,3,5\}$$

The complement is just a 'special minus'; in fact, the complement of a set *A* is u\A.

Notation

The complement of a set *A* is written A^c and is read '*A* complement' or 'the complement of *A*'. In the example above you could have written:

$$D^c = \{2,3,5\}$$

Activity 5.12

Complement 1

Suppose that $\mathsf{U} = \{1,2,3,4,5,6\}$, then what is B^c if $B = \{1,2,3\}$?

Activity 5.13

Complement 2

If the Universal set u is the set of World Cup winners {Brazil, England, Germany, Argentina, Italy} from activity 5.3, and *E* is {England, Germany} as in activity 5.7, then what is E^c?

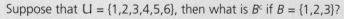

5.8 Extending things

Most of the time when you use sets you will find that you have more than two sets. So, you should first understand how to extend the formation of unions, intersections and minuses to this scenario. For example, we need to understand what we mean by the sets $A \cup B \cup C$ or $A \cap B \cap C$.

In the first chapter of this book you learned how different arithmetic operations, such as ×, + and −, have different priorities in expressions such as:

$2+4 \times 3$.

This expression evaluates to 14; the × has a higher priority than the + so we perform the × first. If we wanted the + to have a higher priority, then we would surround it by brackets:

$(2+4) \times 3$.

In this expression we work out the bracket first (since brackets always have the highest priority), so the expression evaluates to 18.

Example 5.14

Consider the sets:

$B = \{1,2,3,4,5\}$

$C = \{2,4\}$

$D = \{1,2,3\}$.

What are the sets $B \cap C \cap D$ and $B \cup C \cup D$?

The set $B \cap C \cap D$ is the set of elements that are in all of the three sets B, C and D. If you look closely at these sets, then you will see that:

$B \cap C \cap D = \{2\}$

Similarly, the set $B \cup C \cup D$ is the set consisting of all elements in all the sets B, C and D. Again you should be able to see that:

$B \cup C \cup D = \{1,2,3,4,5\}$

Activity 5.14

Multiple intersections and unions 1

If we have the sets:

$A = \{a,b,c\}$

$B = \{b,d,e\}$

$C = \{a,b,e\}$

find the sets $A \cap B \cap C$ and $A \cup B \cup C$.

Activity 5.15

Multiple intersections and unions 2

If we have the sets

W = {Brazil, England, Germany, Argentina, Italy}

X = {England, France, Germany}

Y = {Brazil, England, France, Italy, Peru}

find the sets $W \cap X \cap Y$ and $W \cup X \cup Y$.

When you only have intersections or only have unions, then finding the set is straightforward. However, when you have a mixture, it becomes ambiguous.

Example 5.15

If we have the sets:

A = {a,b,c}

B = {b,d,e}

C = {a,b,e}.

find $A \cap B \cup C$.

Although this example seems fairly straightforward, you will find out that, in fact, it cannot be answered. Look at the following two ways of finding $A \cap B \cup C$.

Suppose that we work out $A \cap B$ first, so

$A \cap B$ = {b}

then

$A \cap B \cup C$ = {b}$\cup C$

= {b}$\cup$\{a,b,e\}

= {a,b,e}.

However, if we work out $B \cup C$ first

$B \cup C$ = {a,b,d,e}

then

$A \cap B \cup C$ = = $A \cap$\{a,b,d,e\}

= {a,b,c}$\cap$\{a,b,d,e\}

= {a,b}.

These two answers are different, which is a potential problem. The reason we get two different answers is precisely because we do things in a different order. For the first answer we got, {a,b,e}, we worked out $A \cap B$ first and then $A \cap B \cup C$. For the second answer, {a,b}, we worked out $B \cup C$ first. But this is exactly like the fact that we could think of 2+4×3 in two different ways, giving us either 14 or 18.

The problem is that unlike the × and + operations, $\cap$ and $\cup$ are given the same priority and so you can't tell which one you should do first. To combat this we use brackets, as in arithmetic. Brackets always have the highest priority, so you work out the contents of the brackets first.

In the example, the first answer should have been written

$(A \cap B) \cup C = \{a,b,e\}$

and the second should have been written

$A \cap (B \cup C) = \{a,b\}$.

Example 5.16

Consider the sets

$B = \{1,2,3,4,5\}$
$C = \{2,4\}$
$D = \{1,2,3\}$.

Find the sets $(B \cup C) \cap D$ and $B \cup (C \cap D)$.

To find $(B \cup C) \cap D$ we first work out the bracket, $(B \cup C)$:

$B \cup C = \{1,2,3,4,5\} \cup \{2,4\}$
$= \{1,2,3,4,5\}$.

So:

$(B \cup C) \cap D = \{1,2,3,4,5\} \cap \{1,2,3\}$
$= \{1,2,3\}$.

Similarly, to find $B \cup (C \cap D)$ we first work out $(C \cap D)$:

$C \cap D = \{2,4\} \cap \{1,2,3\}$
$= \{2\}$.

So:

$B \cup (C \cap D) = \{1,2,3,4,5\} \cup \{2\}$
$= \{1,2,3,4,5\}$.

5.9 Cardinality and the inclusion exclusion law

When you learned about subsets and supersets earlier you found out that if $A \subset B$ then you can think of this as meaning that A is a smaller set than B. This idea is useful to measure the relative size of a set. However, in this section you will find out exactly how to measure how big a set is.

The *cardinality* of a set is the number of elements in that set. In most cases this is easy to find by just adding up the elements. If A is a set, then the cardinality of A is written $\#A$ and is a natural number, $\#A \in \mathbb{N}$, unless A is the empty set. The cardinality of the empty set is 0 because it has no elements.

Example 5.17

Let $A = \{m,a,t\}$ then

$\#A = 3$

since A has three elements.

Activity 5.16

Cardinalities

Write down the cardinalities of the following sets:

$L = \{M, I, S, P\}$

$C = \{red, green, blue\}$

$u = \{1, 2, 3, 4, 5, 6\}$

$W = \{Brazil, England, Germany, Argentina, Italy\}$

So the cardinality of a set is usually easy to calculate. The question is, if we work out the union or intersection of two sets, what is the new cardinality? Let's find out how to work this out.

Example 5.18

Suppose we have the two sets: $A = \{a, d, e, f\}$ and $B = \{c, d, e\}$ find $\#A$, $\#B$ and $\#(A \cup B)$.

Well, $\#A = 4$ and $\#B = 3$. To find $\#A \cup B$ note that:

$$A \cup B = \{a, c, d, e, f\}$$

and so $\#(A \cup B) = 5$.

Is this what you would expect? You have seen that $A \cup B$ is like adding the sets A and B together, so what happens if we add together the cardinalities of A and B? We get 7, which is bigger than $\#(A \cup B)$. So what has happened?

Suppose we write down the elements of the two sets in a table:

$$A = a \; d \; e \; f$$
$$B = c \; d \; e$$

and if we add up all the elements, we get $\#A + \#B$, which is 7. But, when we do this, we are adding up the letters d and e twice, even though in $A \cup B$ the letters d and e only appear once.

So you've found out that if we add up $\#A$ and $\#B$ then we get a number greater than $\#A \cup B$ and it is too big precisely because we have added up some elements twice. The question is, which elements are we adding up twice; in other words, what's so special about the letters d and e?

What is A∩B and what is #(A∩B)?

First $A \cap B = \{d, e\}$ and so $\#(A \cap B) = 2$.

So you've found out from this example that when we add up $\#A$ and $\#B$ we are adding up the elements d and e twice. But these letters are exactly the elements of $A \cap B$.

So, if we want to find $\#(A \cup B)$ then we add up $\#A$ and $\#B$ and subtract $\#(A \cap B)$ (since we have added these elements twice instead of once).

Therefore, you have found the following formula, which is called the *inclusion exclusion law*:

$$\#(A \cup B) = \#A + \#B - \#(A \cap B).$$

The inclusion exclusion law is extremely useful. In chapter 2 you found out that formulas provide a way of working out something from other information. The inclusion exclusion law lets us work out the cardinalities of intersections or unions which can be very important. A typical example is the following.

Example 5.19

A group of students arrives at a class. 12 of them have brought a textbook. 16 of them have brought a calculator. 5 of them have brought both a textbook and a calculator. How many students have brought either a textbook or a calculator (or both)?

Let T be the set of students who have brought a textbook. There are 12 of these students, so $\#T = 12$.

Let C be the set of students who have brought a calculator. There are 16 of these students, so $\#C = 16$.

The set of students who have brought both a textbook and a calculator is the set of students in both T and C, which is the set $T{\cap}C$. There are 5 of these students, so $\#(T{\cap}C) = 5$.

The set of students who have brought either a textbook or a calculator (or both) is the set of students in either T or C (or both), which is the set $T{\cup}C$. We want to find $\#(T{\cup}C)$.

By the inclusion exclusion law we have

$\#(T{\cup}C) = \#T + \#C - \#(T{\cap}C) = 12+16-5 = 23$.

So 23 students have brought either a textbook or a calculator (or both).

Example 5.20

All the students in a certain school either take English or Maths classes; 456 students take English classes and 123 students take Maths classes. There are 60 students who take both English and Maths classes.

How many students are there in this school?

Let E be the set of students who take English and M the set of students who take Maths. The question states that 456 students take English, so $\#E = 456$. Similarly, 123 students take Maths, so $\#M = 123$. We also know that there are 60 students who take both English and Maths. This set is $E{\cap}M$ and so $\#(E{\cap}M) = 60$.

Now we can use the inclusion exclusion law because the total number of students in the school is $\#E{\cup}M$.

$\#(E{\cup}M) = \#E + \#M - \#(E{\cap}M)$

$= 456+123-60$

$= 519$.

The school therefore has 519 students.

Example 5.21

An internet search engine search finds 12,000 websites with the words 'set' or 'theory' on them. If 8,543 websites contain the word 'set' and 10,233 websites contain the word 'theory', how many websites contain both the words?

This example is a little bit different to the last one. If we let S be the set of websites that contain 'set' and T the set of websites that contain 'theory', then we have that $\#S = 8{,}543$ and $\#T = 10{,}233$. The set of websites that contain both the words is $S \cap T$, and the set of websites that the search engine finds is $S \cup T$ and so $\#(S \cup T) = 12{,}000$.

By the inclusion exclusion law we then get:

$\#(S \cup T) = \#S + \#T - \#(S \cap T)$

$12{,}000 = 8{,}543 - 10{,}233 - \#(S \cap T)$. So rearranging:

$\#(S \cup T) = 8{,}543 - 10{,}233 - 12{,}000$

$= 6{,}776.$

Therefore there are 6,776 websites that contain both 'set' and 'theory'.

5.10 Summary

In this chapter we saw how to use the language of sets to say things about collections of objects in an organised way, just as in chapter 1 we used algebra to say things about numbers in an organised way. We saw how to deal with sets that overlap, and how to use the language of sets to solve problems.

5.11 Review questions

 Question 5.1: Defining sets

Write down the set S containing the letters of the word 'MIDDLESEX' (use capital letters).

Question 5.2: Unions

In review question 5.1 you had the set S and in example 5.2 you had the set $L = \{M,I,S,P\}$. Find the set $S \cup L$.

Question 5.3: Intersections

Let the sets, A, B and C be:

$A = \{1,2,3,4,5\}$

$B = \{2,3,4,5,6,7,8\}$

$C = \{-1,0,1,2,3\}.$

Write down the sets $A \cap C$ and $A \cap B$.

Question 5.4: Minus

In review question 5.1 you had the set S and in example 5.2 you had the set $L = \{M,I,S,P\}$. Find the sets $S \backslash L$ and $L \backslash S$.

Question 5.5: Complements

If E is the set of even numbers and $U = \mathbb{N}$ then what is E^c?

 Question 5.6: Sets

Suppose you have the following sets:

$U = \{1,2,3,4,5,6\}$

$A = \{2,3,6\}$

$B = \{1,2,5\}.$

Write down what the following sets are:

1. $A \cup B$

2. $A \backslash B$

3. A^c

4. B^c

5. $A \backslash B^c.$

 Question 5.7: Three sets

If:

$A = \{2,4,6\}$

$B = \{1,2,3,4,5\}$

$C = \{2,4\}.$

Find the sets $(A \cap B) \cup C$ and $(A \cup B) \cap C$.

 Question 5.8: Cardinalities

Suppose that A and B are disjoint sets with $\#A = 7$ and $\#B = 21$. What is the cardinality of $A \cup B$?

 Question 5.9: A practical problem

The staff of a certain company are appraised at the end of each year. Of the members of staff appraised 223 said they were dissatisfied with their working conditions and 367 said they were unhappy about their level of pay. In addition, 144 said that they were both unhappy about their pay and dissatisfied with their working conditions. The human resource manager asks you to find out how many of his staff are unhappy about a part of their job.

Question 5.10: The inclusion exclusion law

A statistician surveys people about their lifestyles. Of those surveyed, 454 said they go to the cinema frequently and 613 said they go to the theatre frequently. The statistician counts up the number of people who either go to the cinema or go to the theatre frequently and gets 1,000. How many of those surveyed said that they go to the theatre and the cinema frequently?

5.12 Answers to review questions

Question 5.1: Defining sets

The solution is:

$S = \{M,I,D,L,E,S,X\}$.

Notice that there is only one D and only one E, since we do not repeat elements in a set.

Question 5.2: Unions

The solution is:

$S \cup L = \{M,I,D,L,E,S,X,P\}$.

Again, we do not repeat elements, so M, I and S appear only once.

Question 5.3: Intersections

The answers are:

$A \cap C = \{1,2,3\}$

$B \cap A = \{2,3,4,5\}$.

Question 5.4: Minus

The answers are:

$S \backslash L = \{D,L,E,X\}$

$L \backslash S = \{P\}$.

Notice that $S \backslash L$ is not the same as $L \backslash S$. Look back at the content of the chapter to see more about this.

Question 5.5: Complements

The set E^c is the set of odd numbers. Sometimes sets cannot be written explicitly but can be described only in terms of the properties of their elements.

Question 5.6: Sets

1. $A \cup B = \{1,2,3,4,5,6\}$
2. $A \backslash B = \{3,6\}$
3. $A^c = \{1,4,5\}$
4. $B^c = \{3,4,6\}$
5. $A^c \backslash B^c = \{1,5\}$

Question 5.7: Three sets

We work out the bracket first $(A \cap B) = \{2,4\}$, then we have

$(A \cap B) \cup C = \{2,4\} \cup C = \{2,4\}$.

For the second one we work out the bracket:

$(A \cup B) = \{1,2,3,4,5,6\}$, then

$(A \cup B) \cap C = \{1,2,3,4,5,6\} \cap C = \{2,4\}$.

When you have brackets in an expression, you must work them out first.

Question 5.8: Cardinalities

This is surprisingly easy. Since A and B are disjoint, we have that $A \cap B = \emptyset$ and so $\#(A \cap B) = 0$. Therefore we get:

$\#(A \cup B) = \#A + \#B - \#(A \cap B)$

$= 7 + 21 - 0$

$= 28$

Question 5.9: A practical problem

Let W be the set of staff that are dissatisfied with their working conditions, then $\#W = 223$. Let P be the set of staff that were unhappy about their pay, then $\#P = 367$. We are also told that $\#P \cap W = 144$, so the total number of staff who are unhappy about a part of their job is:

$\#(P \cup W) = \#P + \#W + \#(P \cap W)$

$= 367 + 223 - 144$

$= 446$

So, 446 staff were unhappy.

Question 5.10: The inclusion exclusion law

Let C be the set of those surveyed that frequent a cinema and T the set of those people that frequent a theatre. Then we have the following facts:

$\#C = 454$

$\#T = 613$

$\#(C \cup T) = 1{,}000$

So, by the inclusion exclusion law we have

$\#(C \cup T) = \#C + \#T - \#(C \cap T)$

$1{,}000 = 454 + 613 - \#(C \cap T)$. Rearranging, we get

$\#(C \cap T) = 454 + 613 - 1{,}000$

$= 67$.

So, only 67 of those surveyed go to both the cinema and the theatre frequently.

5.14 Feedback on activities

Activity 5.1

The set is $A = \{10,11,12,13,14,15,16\}$.

Activity 5.2

You could pick any of the members; let's choose 14. Then we say:

$14 \in A$.

Activity 5.3

The set is

{Brazil, England, Germany, Argentina, Italy}.

Each country is included only once, even if they won the Cup more than once.

Activity 5.4

The obvious way to write the set is

{2000, 2001, 2002, 2003, 2004, 2005}.

However, because the order of the elements doesn't matter, the same set could also be written in many other ways, for example

{2005, 2003, 2000, 2001, 2002, 2004}.

So, if we're checking whether two sets are the same we have to be careful to spot when one is simply a rearrangement of the other, and this can get difficult if the sets are large.

Activity 5.5

The set $D \cup E$ contains all the elements in both the sets. So:

$D \cup E$ = {a,b,c,d,e,f,g}.

Activity 5.6

First we have that the common elements of A and B are $A \cap B$ = {3,4,5}, so:

$A \backslash B$ = {1,2}
$B \backslash A$ = {6,7,8}.

Activity 5.7

$W \backslash E$ = {Brazil, Argentina, Italy}, the set of countries in W but not in E.

$E \backslash W$ = $\emptyset$, the empty set, because there are no countries which are in E but not in W.

Activity 5.8

The list should look like the following:

$A \subset B$	☒	$A \subset C$	☒
$B \subset C$	☒	$B \subset D$	☒
$C \subset A$	☑	$C \subset B$	☑

Activity 5.9

W is not a subset of E, because W contains Brazil (for example) but E does not.

E is a subset of W, i.e. $E \subset W$, because both the elements of E are also in W.

Activity 5.10

As we saw in example 5.2 the set of letters of MISSISSIPPI is {I, S, M P} (we didn't write the letters in this order, but this is the same set).

The set of letters of SIMPLE is {E, I, S, M, L, P}.

Each of the 4 letters in the first set is also in the second set, so the first set is a subset of the second set.

Note that this is true, even though some letters, such as I, appear more often in MISSISSIPPI than they do in SIMPLE. As usual, the number of times an element is counted as being in a set is not important.

Activity 5.11

$\mathbb{N} \subset \mathbb{Z} \subset \mathbb{Q} \subset \mathbb{R}$

- $\mathbb{N} \subset \mathbb{Z}$ – this is true because the integers include all the positive whole numbers and the negative whole numbers and 0.

CHAPTER 6

Introduction to probability

OVERVIEW

In this chapter we will introduce the concept of probability, chance and randomness. We will determine how likely events are to occur when we conduct or observe an experiment. In this chapter we will consider discrete experiments with finite numbers of outcomes.

Learning outcomes On completion of this chapter, you should be able to:
• Define an experiment and its associated outcome set
• Calculate basic discrete probabilities
• Express probabilities as decimals, fractions and percentages
• Demonstrate knowledge of the properties of discrete probabilities.

6.1 Probability and chance

Probability measures the chance that something will happen. Statements about probability occur in everyday speech. For example, the following statements are concerned with chance:

- 'I will probably enjoy this chapter.'
- 'It is highly likely to rain today.'
- 'Nine times out of ten the train I go to work on will be late.'

Probability gives a structure to the idea of chance and allows us to try to measure the level of uncertainty or chance. This will enable us to evaluate the level of associated risk. This level of risk will inform decisions and choices to be made in the future. For example, we could evaluate how risky an investment is and so determine how likely an investment is to make a loss. If I feel that the chance of making a loss is too great, I will not invest.

6.2 Definitions

An *experiment*, *X*, is a situation that can be performed (or considered) to gain information.

Example 6.1

Our experiment could be as simple as picking a student at random from a class list.

Possibly our experiment may be more complicated, for example taking part in a raffle to win one of the prizes, or perhaps playing the lottery.

An *outcome set*, Ω, is a set (list) of possible results associated with an experiment.

Example 6.2

Let *X* = picking a student at random from a class list. Then, Ω is the list of names of all students in the class. For example, Ω = {Androulla, Bal, Clare, Davinder, Elizabeth}.

An *event, E,* is either a single or combination of outcomes.

Example 6.3

Let *X* = picking a student at random from a class list and Ω = {Androulla, Bal, Clare, Colin, Davinder, Elizabeth}.

One event could be picking a student at random from the list and their name beginning with **C**. This condition is satisfied by two of the outcomes (names) in our outcome set, Clare and Colin.

This could be written as, **E** = picking Clare or Colin.

Activity 6.1

Outcomes

State the set of outcomes and one event for the following experiments:

1. Guessing the gender of a baby.

2. Flipping a coin. On one side of the coin there is a face, which we call heads and on the other side is a crest, which we call tails.

3. Throwing a normal 6-sided die.

4. Selecting a chapter from this book.

5. Selecting a country from the European Union.

Example 6.4

Suppose an experiment consists of flipping a coin twice. What is the set of outcomes?

For simplicity let us denote the coin landing 'heads' up by *H*, and 'tails' up by *T*.

The first time we flip the coin it could land either 'heads' up or 'tails' up: two results.

The second time we flip the coin it could also land either 'heads' up or 'tails' up: two results.

So, if the first result is a 'heads' then the result of the second flip will be one of two possible answers: *H* or *T*, and if the result of the first flip is 'tails' then the result of the second flip will also be one of two possible answers. This gives four (2x2) outcomes to list.

It is much easier to list these in a table.

Table 6.1: Example 6.4 outcomes

		Second result	
		H	T
First result	H	(H, H)	(H, T)
	T	(T, H)	(T, T)

So, $\Omega = \{ (H, H), (H, T), (T, H), (T, T) \}$.

Activity 6.2

Drawing balls

Suppose that a bag contains three balls; one red, one blue and one green. We pick a ball from the bag, note its colour and then return it to the bag. This process is then repeated a second time.

List all nine outcomes for this experiment.

		Second result		
		Red	Blue	Green
First result	Red			
	Blue			
	Green			

Activity 6.3

Flipping coins

1. Suppose we flip a coin three times. What are the possible outcomes? Hint: in this case it's difficult to lay the possibilities out in a table so it's probably better simply to list them. Try to list them in a logical order, however.

2. Suppose we flip a coin four times. How many possible outcomes are there?

6.3 A scale of probability

Probabilities are measured on a scale between 0 and 1. Sometimes we refer to the percentage chance. We convert our probability value that is between zero and one to a percentage, as we did in chapter 4. Hence our percentage chance will lie between 0% and 100%.

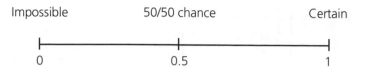

If the probability of an event, $P(E)$, is 0, then the event is impossible. For example, the probability that the following events will occur is 0 since they are all impossible:

- The probability of a person being 100 metres tall
- The probability that I select a black ball from an empty bag
- The probability of winning a raffle if I do not have a ticket.

If the probability of an event, $P(E)$, is 1, then the event is certain. For example, the probability that the following events will occur is 1 since they are all certain to occur:

- The probability of a person being less than 100 metres tall
- The probability that I select a black ball from a bag that contains only black coloured balls
- The probability of winning a raffle if I have all of the tickets.

The sum of all probabilities associated with an experiment is 1.

6.4 Evaluating probabilities using the classical method

We will discuss two methods for evaluating probabilities in this chapter: classical and observational.

The *classical method* assumes that all of the events in the outcome set for an experiment are equally likely. The *observational method* estimates probabilities by repeated experimentation and observation.

The classical method

This method assumes that the experiment has a finite number of outcomes that are equally likely.

Definition

For an event E associated with an experiment X, the probability of observing the event is denoted by $P(E)$ and is defined as the following:

$$P(E) = \frac{\text{Number of ways an event can occur}}{\text{Total number of outcomes for the experiment}}$$

Example 6.5

Suppose that an experiment, X, consists of rolling a six-sided die, and noting the number of spots showing on the top face. See below.

Figure 6.1: Six-sided dice

Then the set of outcomes $\Omega = \{1, 2, 3, 4, 5, 6\}$.

Now, if we assume that the die is fair, so perfectly balanced, then each face of the die is equally likely to be rolled. In fact, each face has a one-in-six chance of being rolled.

So, $P(1) = P(2) = P(3) = P(4) = P(5) = P(6) = \frac{1}{6}$.

Let our event E be obtaining an even number. So E is obtaining 2 or 4 or 6, which are three of our six outcomes. Hence the probability $P(E)$ that the die will result in an even number is:

$$P(E) = \frac{Number\ of\ ways\ an\ event\ can\ occur}{Total\ number\ of\ outcomes\ for\ the\ experiment} = \frac{3}{6} = \frac{1}{2}.$$

Activity 6.4

Rolling a die

Suppose that an experiment, X, consists of rolling a six-sided fair die as above and noting the number of spots showing on the top face. So $\Omega = \{1, 2, 3, 4, 5, 6\}$.

Calculate the following probabilities,

1. The result is an odd number.

2. The result is less than 5.

3. The result is 7.

Example 6.6

A standard pack of 52 playing cards consists of 4 suits: hearts ♥, diamonds ♦, clubs ♣ and spades ♠.

Each suit consists of 13 cards, an ace, numbered cards from 2 to 10 and 3 face cards which are a jack, a queen and a king. See figure 6.2.

Figure 6.2: Playing cards

Suppose that the pack is thoroughly shuffled and a card is selected from the pack at random. What is the probability that the card selected is:

1. The king of diamonds?
2. A seven of any suit?
3. A club?
4. An even-numbered card?

There are 52 cards in the pack and they are equally likely to be picked. Hence, every card has a 1 in 52 chance of being selected.

1. There is only one king of diamonds in the pack, and so:

   $P \text{ (king of diamonds)} = \frac{1}{52}$.

2. There are four sevens in the pack, one in each of the four suits, so:

   $P \text{ (seven of any suit)} = \frac{4}{52} = \frac{1}{13}$.

3. There are 13 clubs contained in the pack so:

   $P \text{ (club)} = \frac{13}{52} = \frac{1}{4}$.

4. To calculate this probability you must work out how many even cards are contained in any suit and then multiply it by 4. It can be helpful to visualise this.

Table 6.2: Even-numbered playing cards

	Ace	2	3	4	5	6	7	8	9	10	Jack	Qn	King
♥		✓		✓		✓		✓		✓			
♣		✓		✓		✓		✓		✓			
♦		✓		✓		✓		✓		✓			
♠		✓		✓		✓		✓		✓			

We can see there are five even-numbered cards in each of the four suits, hence we have:

$P \text{(even-numbered card)} = \frac{20}{52} = \frac{5}{13}$.

Activity 6.5

Playing cards

Let us use a standard pack of 52 playing cards comprising four suits: hearts ♥, diamonds ♦, clubs ♣, and spades ♠, with each suit consisting of 13 cards, an ace, numbered cards from 2 to 10 and 3 face cards which are a jack, a queen and a king.

Suppose that a card is selected from the pack at random from a thoroughly shuffled pack. What is the probability that the card selected is:

1. The queen of hearts?

2. A numbered card which is a multiple of 3 or 5?

Express the probabilities as decimals and percentages.

Activity 6.6

Random email

A faulty wireless connection means that an email consists of random letters instead of the original message. If each letter of the alphabet is equally likely, what is the probability that the first letter of the email is a vowel (A, E, I, O or U)?

Activity 6.7

Chapters

If you select one of the chapters of this book at random, what is the probability that it will be either one of the two on summary statistics or the one on sets?

6.5 Evaluating probabilities using the observational method

Sometimes it will not be possible to use the classical approach to work out probabilities. It will be necessary to estimate probabilities using data we have observed from conducting an experiment.

For example, we could estimate the probability that a student in 2002 passes the end-of-chapter self-assessment by finding out how many students have passed during the previous three years and using this proportion as an estimate of the probability that a student will pass the end-of-chapter self-assessment in 2002.

It would not be possible to use the classical method here as we do not have a theoretical value for the probability that someone will pass the test.

Example 6.7

Table 6.3 details the number of students who have passed the end-of-chapter self-assessment for the three years prior to 2002.

Table 6.3: Number of students passing end-of-chapter self-assessment

	Number of students	
	Pass	Fail
1999	200	10
2000	500	20
2001	750	25

Based on this information, how likely would it have been for a student to pass the end-of-chapter self-assessment in 2002?

First we must work out how many students have passed the test in total over the three years. To do this we add all the entries in the pass column. Then we must find out how many students have taken the test, either passing or failing, in total. To do this we add all the entries in the table. This is then expressed as a proportion.

Table 6.4: Total numbers of students taking the end-of-chapter self-assessment

	Number of students		
	Pass	Fail	Totals
1999	200	10	210
2000	500	20	520
2001	750	25	775
Totals	1,450	55	1,505

So, based on this information, the probability that a student will pass the end-of-chapter self-assessment is:

$$\frac{Total\ no.of\ students\ to\ pass}{Total\ no.\ of\ students} = \frac{1,450}{1,505} = 0.9634 = 96.34\%$$

Activity 6.8

A biased die?

You observe a die being rolled and suspect that it is not balanced and is biased in favour of one of the faces. You roll the die 120 times and observe the following:

Result of top face	1	2	3	4	5	6
Frequency	9	33	39	27	7	5

1. Estimate the probabilities below by completing the table.

Result of top face E	1	2	3	4	5	6
Probability of result $P(E)$						

2. Estimate the probability that the result will be a 2 or 3 or 4.

Example 6.8

A firm collects information concerning the punctuality of staff in different departments. The results are displayed in the following table.

	Level of punctuality		
	Early	On time	Late
Department A	38	20	10
B	18	115	60
C	4	35	100

How likely are the following?

1. A member of staff being on time for work.
2. A member of staff being late and in department B.
3. A member of staff being in department A.

In order to estimate the probabilities, we will need to evaluate the totals. It will then be possible to express these as proportions.

	Level of punctuality			
	Early	On time	Late	Totals
Department A	38	20	10	68
B	18	115	60	193
C	4	35	100	139
Total	60	170	170	400

1. If we look in our table, there are 170 members of staff who are on time for work out of a total of 400 who work for the firm. Hence we can estimate the probability as $\frac{170}{400}$.

2. There are 60 members of staff who are late and are in department B out of a total of 400. Hence we can estimate the probability as $\frac{60}{400}$.

3. There are 68 members of staff who work in department A out of a total of 400. Hence we can estimate the probability as $\frac{68}{400}$.

Staff punctuality

Use the information concerning the punctuality of staff in different departments from example 6.8 displayed in the following table to estimate the following probabilities:

| | | Level of punctuality | | |
		Early	On time	Late
Department	A	38	20	10
	B	18	115	60
	C	4	35	100

1. A member of staff being early or on time for work.

2. A member of staff being in department B or C.

Laptop survey

A computer magazine is conducting a survey of popular laptop makes: *L*, *M* and *N*. They interview passengers with laptops at an airport and ask them how satisfied they are with their laptop. The results are as follows:

| | | Satisfaction | | |
		Very satisfied	Quite satisfied	Not satisfied
Laptop	*L*	8	9	12
	M	12	18	6
	N	6	8	2

Use this information to find each of the following probabilities:

1. A person has laptop *M*.

2. A person is very satisfied with their laptop.

3. A person has laptop *L* but they are not satisfied with it.

Activity 6.9

Activity 6.10

Activity 6.11

Complaints log

A computer support company logs complaints from clients. For each complaint. it records which package the client is using and which make of computer it is running on. The results are as follows:

		Make of computer		
	X	Y	Z
Package A	23	20	5
B	5	23	7
C	18	28	2
D	10	25	5

Use this information to find each of the following probabilities:

1. A client is using computer X.

2. A client is using package A or package B.

3. A client has a problem with package C on computer Y.

4. A client has a problem with computer Y but is not using package B.

6.6 Summary

In this chapter we saw what is meant by probability. We saw how to calculate probabilities, and how to use probability to solve problems. Although many of the examples were about gambling games, there are lots of real-world situations that can be thought of using probability ideas.

6.7 Review questions

Question 6.1: Outcomes

State the set of outcomes and one event for the following experiments, X:

1. X = Picking a chapter at random from this book to study.

2. X = Picking a member of a household at random.

Question 6.2: Mystery prizes

Suppose that you are selected to pick two prizes at random. The rules state that you must pick one prize from bag A and one prize from bag B.

Bag A contains a box of chocolates and a box of jelly beans.

Bag B contains a can of cola drink, a can of orange juice and a can of lemonade.

List all six outcomes for this experiment:

		Bag B		
		Cola	Orange juice	Lemonade
Bag A	Chocolates			
	Jelly beans			

So, $\Omega = \{$ $\}$.

Question 6.3: Outcomes

Suppose that a bag contains four balls: one red, one blue, one green and one orange. An experiment consists of picking a coloured ball from the bag twice after replacing the first ball picked and noting the result each time.

List all 16 outcomes for this experiment.

		Second result			
		Red	Blue	Green	Orange
First result	Red				
	Blue				
	Green				
	Orange				

So, $\Omega = \{$ $\}$.

Question 6.4: Rolling a die

Suppose that an experiment, X, consists of rolling a six-sided fair die as in example 6.5, and noting the number of spots showing on the top face.

1. State the set of outcomes for this experiment.

Calculate the following probabilities:

2. The result is less than 6.

3. The result is 1 or more.

4. P (1 or 3).

5. P (number is divisible by 3).

6. P (prime).

Question 6.5: Card games

Let us use a standard pack of 52 playing cards comprising four suits: hearts ♥, diamonds ♦, clubs ♣, and spades ♠, with each suit consisting of 13 cards, an ace, numbered cards from 2 to 10, and 3 face cards which are a jack, a queen and a king.

Suppose that a card is selected at random from a thoroughly shuffled pack. What is the probability that the card selected is:

1. A face card or an ace?

2. A numbered card from a red suit, so ♥ or ♦?

Question 6.6: Rolling a die 2

A die is rolled 100 times. The frequency of each result is given in the table below:

Result of top face, E	1	2	3	4	5	6
Frequency	14	18	16	14	18	20

Using this information, estimate the probability of each of the following results:

Result of top face, E	1	2	3	4	5	6
Probability of result, P(E)						

Question 6.7: Rolling a die 3

This activity requires you to make a die from card. Use the following template as a guide to construct your die. Make sure that each side of the die is square. Once you have drawn out your template onto the card, cut it out and fold the sides up and over, making a cube. Stick this together using adhesive tape.

1. Roll your die 30 times and note the results in the table which follows.

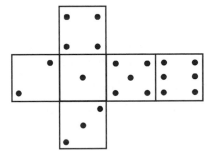

Result of top face, E	1	2	3	4	5	6
Frequency						

2. Estimate the probabilities of the following events:

Result of top face, E	1	2	3	4	5	6
Probability of result P(E)						

3. Looking at the values you have obtained, do you believe that your die is *fair* or *biased*?

4. Repeat parts 1 and 2 with your die, this time rolling it an additional 30 times. Now that you have information for 60 rolls of the die, do you still hold the same belief that the die is fair or biased?

Question 6.8: Defective components

A random sample of 309 electronic component defects were recorded and classified according to the type of defect (A, B, C, D) and the supplier (X, Y, Z) they were manufactured by.

The following table displays the supplier against the type of defect.

		Type of defect			
		A	B	C	D
Supplier	X	15	21	45	13
	Y	26	31	34	5
	Z	33	17	49	20

Estimate the following probabilities:

1. A defective component is manufactured by supplier X.

2. The type of defect found is classified as type B.

3. A defective component was manufactured by supplier Y and classified as type C.

Question 6.9: Meal expectations

A food critic notes the price of a meal and whether the meal was good, very good, or excellent on 300 occasions.

		Cost of meal			
		£10	£15	£20	£25
Quality	Good	42	40	2	0
	Very good	34	64	46	6
	Excellent	2	14	28	22

Estimate the probabilities that the next meal the food critic has will be:

- Very good or excellent

- More than £10

- A good meal and cost £10.

6.8 Answers to review questions

Question 6.1: Outcomes

1. X = Picking a chapter at random from this book to study. Then **W** is the list of the chapters in this module. So your outcome set will be:

   Ω = {chapter 1, chapter 2, chapter 3, chapter 4, chapter 5, chapter 6, chapter 7, chapter 8, chapter 9, chapter 10}.

   One possible event is picking a chapter with an even number, so it would be picking one of chapters 2, 4, 6, 8 and 10.

2. X = Picking a member of household. Then Ω is the list of everybody who shares a house with you. So your outcome set will probably differ from those of your classmates.

   It may look like this:

   Ω = {mother, father, older brother, myself, younger sister, grandparent}

   One event could be picking a member of your household who is older than you. Another could be picking one who is female.

Question 6.2: Mystery prizes

Now, Bag A contains a box of chocolates and a box of jelly beans, which is two results.

Bag B contains a can of cola drink, a can of orange juice and a can of lemonade which is three results.

As Bags A and B are independent, there are 2x3 possible outcomes to this experiment.

The list of all six outcomes for this experiment is as follows:

		Bag B		
		Cola	Orange juice	Lemonade
Bag A	Chocolates	(Choc, cola)	(Choc, OJ)	(Choc, lem)
	Jelly beans	(JBs, cola)	(JBs, OJ)	(JBs, lem)

So, Ω = {(Chocolates, cola), (Chocolates, orange juice), (Chocolates, lemonade), (Jelly beans, cola), (Jelly beans, orange juice),(Jelly beans, lemonade)}.

Question 6.3: Outcomes

For simplicity let us denote the colour of the ball picked each time in the following way: red by R, blue by B, green by G and orange by O.

The first time we pick a ball it could be coloured either R, B, G or O: 4 results. The second time we pick a ball it could also be coloured either R, B, G or O. So there are 4x4 possible answers. These are listed below.

		Second result			
		Red	Blue	Green	Orange
	Red	(R, R)	(R, B)	(R, G)	(R, O)
	Blue	(B, R)	(B, B)	(B, G)	(B , O)
First result	Green	(G, R)	(G, B)	(G, G)	(G, O)
	Orange	(O, R)	(O, B)	(O, G)	(O, O)

So, Ω = {(R, R), (R, B), (R, G), (R , O), (B, R), (B, B), (B, G), (B, O), (G, R), (G, B), (G, G), (G, O), (O, R), (O, B), (O, G), (O, O) }.

Question 6.4: Rolling a die 1

Let x = rolling a six-sided fair die. Then:

1. The set of outcomes for this experiment is
   Ω = { 1, 2, 3, 4, 5, 6 }

2. P (less than 6) = P (1 or 2 or 3 or 4 or 5) = $\frac{5}{6}$

3. P (1 or more) = P (1 or 2 or 3 or 4 or 5 or 6) = 1

4. $P(1 \text{ or } 3) = \frac{2}{6} = \frac{1}{3}$.

5. $P(\text{number is divisible by } 3) = P(3 \text{ or } 6) = \frac{2}{6} = \frac{1}{3}$.

6. $P(\text{prime}) = P(2 \text{ or } 3 \text{ or } 5) = \frac{3}{6} = \frac{1}{2}$.

Question 6.5: Card games

There are 52 cards in the pack and they are equally likely to be picked. Hence every card has a 1 in 52 chance of being selected.

1. A face card or an ace:

	Ace	2	3	4	5	6	7	8	9	10	Jack	Qn	King
♥	✓										✓	✓	✓
♣	✓										✓	✓	✓
♦	✓										✓	✓	✓
♠	✓										✓	✓	✓

We can see there are four such cards in each of the four suits, hence we have:

$P(\text{face card or an ace}) = \frac{16}{52} = \frac{4}{13}$.

2. A numbered card from a red suit:

	Ace	2	3	4	5	6	7	8	9	10	Jack	Qn	King
♥		✓	✓	✓	✓	✓	✓	✓	✓	✓			
♣													
♦		✓	✓	✓	✓	✓	✓	✓	✓	✓			
♠													

We can see there are nine numbered cards in each of the two red suits, hence we have:

$P(\text{numbered card from a red suit}) = \frac{18}{52} = \frac{9}{26}$.

Question 6.6: Rolling a die 2

To estimate the probabilities, express the frequency of each result as a proportion of the total number of rolls of the die. In this case the die was rolled 100 times. Estimated probabilities are as follows:

Result of top face, E	1	2	3	4	5	6
Frequency	14	18	16	14	18	20
Probability of result $P(E)$	$\frac{14}{100}$	$\frac{18}{100}$	$\frac{16}{100}$	$\frac{14}{100}$	$\frac{18}{100}$	$\frac{20}{100}$

Question 6.7: Rolling a die 3

Parts 1 and 2:
Suppose you observe the following frequencies. Then you must make sure that the sum of the frequencies is 30, and then express the frequencies as proportions out of 30 to estimate the probabilities.

Result of top face, E	1	2	3	4	5	6
Frequency	a	b	c	d	e	f
Probability of result $P(E)$	$\frac{a}{30}$	$\frac{b}{30}$	$\frac{c}{30}$	$\frac{d}{30}$	$\frac{e}{30}$	$\frac{f}{30}$

3. If your die is fair, then you would expect to see a frequency of roughly five in for each value. However, we have only rolled the die 30 times. This is not large so we really do not have enough information in order for us to determine whether the die is fair or biased.

4. Repeating the experiment another 30 times produces new frequencies that total 60. These new frequencies are expressed as proportions out of 60 to estimate the probabilities. We now have more information to help us determine whether the die is biased. However, we still may not be certain that the die is fair or biased as 60 is still not a very large number. If we want to have a higher level of confidence in our determination, we would need to continue rolling the die several hundred times.

Question 6.8: Defective components

To estimate the probabilities, we express frequencies as proportions of the total number of components.

		Type of defect				
		A	B	C	D	Totals
Supplier	X	15	21	45	13	94
	Y	26	31	34	5	96
	Z	33	17	49	20	119
	Totals	74	69	128	38	309

1. The number of defective components manufactured by supplier X is 94. Hence P (defective component is manufactured by supplier X) $= \frac{94}{309}$.

2. The number of type B defects is 69. Hence P (type of defect found is classified as type B) $= \frac{69}{309}$.

3. The number of defective components manufactured by supplier *Y* and classified as type C is 34. Hence

P (defective component is manufactured by supplier *Y* and classified as type C) = $\frac{34}{309}$.

Question 6.9: Meal expectations

Find the required totals and express them as proportions of the 300 meals in total.

		Cost of meal				
		£10	£15	£20	£25	Totals
Quality	Good	42	40	2	0	84
	Very good	34	64	46	6	150
	Excellent	2	14	28	22	66
	Totals	78	118	76	28	300

The number of very good or excellent meals is 150+66 =216.

So we estimate *P* (very good or excellent) = $\frac{216}{300}$.

The number of meals that cost more than £10 is 118+76+28 = 222.

So we estimate *P* (cost more than £10) = $\frac{222}{300}$.

The number of meals that are classified as good and that cost £10 is 42.

So we estimate *P* (good meal and costs £10) = $\frac{42}{300}$.

6.9 Feedback on activities

Activity 6.1

1. Let *X* = guessing the gender of a baby and
   Ω = {boy, girl}.
   One event could be the mother giving birth to a girl.

2. Let *X* = flipping a coin and Ω = {heads, tails}.
   One event could be the coin landing heads up.

3. Ω = {1, 2, 3, 4, 5, 6}
   One event could be scoring a 3.

4. Ω = {Chapter 1, chapter 2, …, chapter 10}
   One event could be selecting chapter 6.

5. There are 25 countries in the European Union (in 2006), so Ω is the set of these 25 countries. One event could be selecting the United Kingdom.

Activity 6.2

For simplicity, let us denote the colour of the ball picked each time in the following way: Red by R, Blue by B, and Green by G.

The first time we pick a ball it could be coloured either R, B or G, 3 results. The second time we pick a ball it could also be coloured either R, B, or G. So there are 3x3 possible answers. These are listed below.

		Second result		
		Red	Blue	Green
	Red	(R, R)	(R, B)	(R, G)
First result	Blue	(B, R)	(B, B)	(B, G)
	Green	(G, R)	(G, B)	(G, G)

So, Ω = {(R, R), (R, B), (R, G), (B, R), (B, B), (B, G), (G, R), (G, B), (G, G)}

Activity 6.3

1. The first flip has two possible outcomes, H or T. For each of these the second flip has two possible outcomes, H or T. So far we have 2x2 = 4 possible outcomes. For each of these the third flip has two possible outcomes, H or T. Overall we therefore have 4x2 = 8 possible outcomes. These are

   Ω = {HHH, HHT, HTH, HTT, THH, THT, TTH, TTT}

2. As before the first flip has two possible outcomes, H or T, and each subsequent flip multiplies the number of outcomes so far by two. So, with four flips the overall number of outcomes is

   $2 \times 2 \times 2 \times 2 = 2^4 = 16$

   We weren't asked to list all these outcomes, but as a check here they are:

   Ω = {HHHH, HHHT, HHTH, HHTT, HTHH, HTHT, HTTH, HTTT, THHH, THHT, THTH, THTT, TTHH, TTHT, TTTH, TTTT}

Activity 6.4

We have X = rolling a six-sided fair die, and Ω = { 1, 2, 3, 4, 5, 6 }.

Since the die is fair, we can assume that each outcome is equally likely to occur. Using the classical probability definition, we obtain:

1. P (an odd number) = P (1 or 3 or 5) = $\frac{3}{6}$ =0.5
2. P (the result is less than 5) = P (1 or 2 or 3 or 4) = $\frac{4}{6} = \frac{2}{3}$
3. P (the result is 7) = 0.

Activity 6.5

There are 52 cards in the pack and they are equally likely to be picked. Hence every card has a 1 in 52 chance of being selected.

1. There is only one queen of hearts in the pack and so

P (queen of hearts) $= \frac{1}{52} = 0.019 = 1.9\%$.

2. In order to calculate this probability we must determine how many numbered cards contained in the pack are multiples of three, ✓, or five, ✔.

Table 6.2: Odd-numbered playing cards

	Ace	2	3	4	5	6	7	8	9	10	Jack	Qn	King
♥			✓		✔	✓			✓	✔			
♣			✓		✔	✓			✓	✔			
♦			✓		✔	✓			✓	✔			
♠			✓		✔	✓			✓	✔			

We can see that there are five cards which are either a multiple of three or five in each of the four suits, hence we have:

P (numbered card multiple of three or five) $= \frac{20}{52} = \frac{5}{13} = 0.385 = 38.5\%$.

Activity 6.6

There are 26 letters in the alphabet and each of them is equally likely to appear. Five of them are vowels. So

P (vowel) $= \frac{5}{26} = 0.1923 = 19.23\%$.

Activity 6.7

There are ten chapters in the book. We're asked for the *single* probability that it will be one of three specified chapter (we're *not* being asked to consider the summary statistics chapters separately from the sets chapter). Assuming that each chapter is equally likely, the probability is:

P (specified chapter) $= \frac{3}{10} = 0.3 = 30\%$.

Activity 6.8

1. Estimate all of the probabilities below by completing the table.

Result of top face, E	1	2	3	4	5	6
Frequency	9	33	39	27	7	5
Probability of result $P(E)$	$\frac{9}{120}$	$\frac{33}{120}$	$\frac{39}{120}$	$\frac{27}{120}$	$\frac{7}{120}$	$\frac{5}{120}$

2. Estimate the probability that the result will be a 2 or 3 or 4.

$$P(2 \text{ or } 3 \text{ or } 4) = \frac{99}{120}$$

Activity 6.9

In order to estimate the probabilities, we will need to evaluate the totals. It will then be possible to express these as proportions.

		Level of punctuality		
	Early	On time	Late	Totals
Department A	38	20	10	68
B	18	115	60	193
C	4	35	100	139
	60	170	170	400

1. There are 230 members of staff who are early or on time for work out of a total of 400.

Hence we can estimate the probability as $\frac{230}{400}$.

2. There are 332 members of staff who are in department B or C out of a total of 400.

Hence we can estimate the probability as $\frac{332}{400}$.

Activity 6.10

A computer magazine is conducting a survey of popular laptop makes: *L*, *M* and *N*. They interview passengers with laptops at an airport and ask them how satisfied they are with their laptop. The results are as follows:

		Satisfaction			Total
		Very satisfied	Quite satisfied	Not satisfied	
Laptop	*L*	8	9	12	29
	M	12	18	6	36
	N	6	8	2	16
	Total	26	35	20	81

82 people were interviewed in total.

1. 36 people have laptop M, so the probability is $\dfrac{36}{81} = \dfrac{12}{27}$

2. 26 people are very satisfied with their laptop, so the probability is $\dfrac{26}{81}$

3. 12 people have laptop *L* but are not satisfied with it, so the probability is $\dfrac{12}{81} = \dfrac{4}{27}$

Activity 6.11

As before we calculate the row and column totals:

		Make of computer			Total
		X	*Y*	*Z*	
Package	*A*	23	20	5	48
	B	5	23	7	35
	C	18	28	2	48
	D	10	25	5	40
	Total	56	96	19	171

171 clients were recorded in total.

1. 56 clients use computer *X*, so the probability is $\dfrac{56}{171}$

2. 48 clients use package *A* and 35 use package *B*, so the probability is $\dfrac{83}{171}$

3. 28 clients use package *C* on computer *Y*, so the probability is $\dfrac{28}{171}$

4. 96 clients have a problem with computer *Y*, but 23 of these are using package *B*; so the number of these *not* using package *B* is 96 − 23 = 73, so the probability is $\dfrac{73}{171}$

CHAPTER 7

Summarising and presenting data

OVERVIEW

Much of the information we need to process as consumers, in the workplace and in our everyday lives comes in the form of numbers, graphs and tables. This chapter will introduce you to and give you a working knowledge of some of the most popular methods used for summarising and presenting data.

The chapter begins with a brief discussion of different types of data and then moves on to discuss the graphical and tabular methods suitable for summarising these different data types. The chapter will conclude with giving you ideas of how this topic extends to other areas that are not formally covered in this book.

Learning outcomes On completion of this chapter, you should be able to:

- Identify different types of data

- Use simple frequency tables to summarise information in a data set

- Use percentages to draw sensible conclusions from data presented in a frequency table

- Interpret data presented graphically in the form of a bar chart or histogram

- Construct and interpret a stem and leaf plot.

7.1 Types of data

Data is a term you will come across repeatedly in your study, in the workplace or in just reading or listening to news articles. What does it actually mean? People tend to think of data as collections of numbers. However, data is not always numerical and even numerical data can be of many different types. Data is simply a scientific term for facts, figures, information and measurements. The type of data that you have will determine what sort of statistical analysis is appropriate, so it is important to determine at the beginning what sort of data you have.

Activity 7.1

Data relating to you

Consider the following list of questions that are all aimed at finding things out about you. Write down your answer to each of the questions.

1. How old are you?

2. Are you male or female?

3. Where were you born?

4. How good do you think you are at mathematics: excellent, very good, average, below average or poor?

5. How many brothers and sisters do you have?

6. Which activities in this book have you completed?

7. What is your favourite food?

8. How many of these answers were given in the form of a number? How many in the form of a set of numbers? How many in some other form?

Variables

Now suppose you were in a class of 30 students, all of whom were asked to complete activity 7.1. The answers for different students would probably be different – each answer would vary from one student to another, so is said to be a *variable*. For example, if we were to write the ages of all 30 students down in a list, this would be a variable that tells us something about the ages of students in the class.

Variables can be of two main types:

Categorical (qualitative) variables

A categorical variable is one where the values are *attributes* which divide the people or objects of the data set into groups. Each person or object must fall into one, *and only one,* of these categories. For example, a person's gender (male or female) and country of birth (Britain, America, China, France etc.) are all category data.

Sometimes the categories have a natural ordering, in which case the variable is said to be *ordinal*. Consider the question in activity 7.1 that asked how good you were at mathematics. The data collected from this question is categorical, as you have to answer in one of five groups: excellent, very good, average, below average or poor. Somebody who answers excellent is probably better at mathematics than someone who answered average.

(However, in saying that, you do need to be careful; this sort of statement assumes everyone answered honestly. It also assumes that everyone has the same idea about what makes someone excellent or average at mathematics, which is probably not true.) Therefore, this is an ordinal variable. It has five categories, and excellent is better than very good etc.

Where there is no such natural ordering, the variable is said to be *nominal*, indicating that its values are just names in categories. A person's gender (male or female) and country of birth are both nominal variables.

Numerical variables

As the heading suggests, these are variables whose values are numbers. Numerical variables can be either *discrete* or *continuous*.

A *discrete variable* is one that can be measured precisely or is restricted to isolated values. The most usual kind of discrete numerical variable is a count, representing the number of something, with possible value 0, 1, 2 etc. For example, the number of brothers and sisters a person has is an example of a variable that will have discrete numerical values.

A *continuous variable* can, in principle, take any value and is typically measured rather than counted. Standard measurements, like lengths, weights and heights, are all examples of continuous variables. In theory all of these things could be measured for a given situation and recorded with a never-ending amount of decimal places. In practice, they will be measured to a certain degree of precision, such as measuring someone's height in centimetres to the nearest two decimal places.

For the purpose of analysis, very large counts such as the population of a country are usually treated as if they were continuous. Similarly, monetary variables such as a person's income or a company's profits are usually thought of as continuous, although they are really discrete.

Sometimes a numerical variable will be *categorised*; for example a person's age in years might be grouped into categories 0-4, 5-9, 10-14 etc. Maybe the categories will not specify the range of ages at all and you may use someone's age to categorise them as young, middle-aged or old, forming an ordinal variable.

Activity 7.2

Identifying types of data

Consider the following variables. Decide whether the type of data for each one is categorical, discrete numerical or continuous numerical. If you decide it is a categorical variable, explain whether it is nominal or ordinal data.

1. The number of graphs in this chapter.

2. The colour of someone's eyes.

3. Opinions expressed on a questionnaire – from strongly disagree, disagree, don't know, agree, to strongly agree.

4. The height of the tallest building in capital cities around the world.

5. The number of passengers passing through an airport on a given day.

Activity 7.3

Data types

Find some data in a newspaper or in a magazine, or perhaps a television report or on a web page. What kind of variables have been measured or observed: categorical or numerical etc?

7.2 Summarising categorical data

Variables can contain a lot of information. For example, in the previous section we said that if we put questions to 30 students we could get 30 different lists of answers. With a list of this size, we could probably just about manage to read down the list for each question and be able to make judgements about common answers, the range of answers, answers which looked totally different to all the others, etc. However, if we had asked the same questions of 2,000 students we would find it impossible to read down the list of answers and make any meaningful sense of it. Therefore, we need to find ways of presenting the information in a concise and meaningful manner so that patterns and characteristics in the variable are immediately apparent. Tables and graphs are a very effective way of summarising and presenting the information in a variable or data set.

Frequency tables

One of the easiest and most effective ways of presenting data is in a table. This is perhaps the most widely used method of data presentation. Whenever you pick up a newspaper, magazine or report, you are likely to see a table. Spreadsheets make the design and manipulation of tables very easy.

A categorical variable is usually summarised in a table of counts, which can, in turn, be summarised by expressing them as percentages of the total:

$$\text{percentage} = \frac{\text{count}}{\text{total}} \times 100\%.$$

It is usually enough to express a percentage to the nearest whole number.

Example 7.1

Consider the following data set which details the gender (male or female) of 15 students in a class.

male   female female male   male   male   male   female female male
male   male   female male   female

Table 7.1 shows a frequency table of this variable.

Table 7.1: Gender of students in a class

Gender	Count	%
Female	6	$\frac{6}{15} \times 100 = 40\%$
Male	9	$\frac{9}{15} \times 100 = 60\%$
Total	15	100%

When you've constructed a frequency table, it's important to check that the frequencies add up to the right number. So, in Example 7.1 we were told that there were 15 students in the class, so we should check that the counts in the table add up to 15. We should also check that the percentages add up to 100% (although the total might be slightly different due to rounding errors).

Activity 7.4

Creating a frequency table 1

Consider the following data set which describes the gender of the students in another class of 30 students:

female	male	male	female	female	male	male	male	male
male	male	male	female	male	male	male	female	male
male	male	male	male	male	male	male	male	male
male	male	female						

Create a frequency table of the gender of students in this second class. Is it true to say that there is the same amount of female students in both classes?

Activity 7.5

Creating a frequency table 2

Consider the following data set which describes the operating system being used on laptops in a sample of people in an airport:

Windows	MacOS	Windows	Windows	MacOS
Windows	Windows	MacOS	Linux	Windows
MacOS	MacOS	Windows	Windows	MacOS
Windows	Windows	Windows	MacOS	Linux

Create a frequency table of the operating system choice in this sample. Which operating system is most popular? Which is least popular?

Graphical methods

Instead of presenting data in a table, it might be better to give a visual display in the form of a graph or chart. Visual displays are good for summarising the data and drawing attention to a particular point. They can also be useful for comparing data sets. Tables, on the other hand, usually give more detailed information about the data set.

The success of any presentation can be judged by how easy it is to understand. A good presentation should make information clear and allow you to see the overall picture. But, good presentations do not happen by chance and need careful planning. If you look at a diagram or table and cannot understand it, it is most probable that the presentation is poor and the fault is with the presenter rather than the viewer.

Sometimes, even when a presentation seems clear, you can look closer and see that it does not give a true picture of the data. This may be a result of poor presentation but sometimes comes from a deliberate decision to present data in a form that is misleading. The problem is that diagrams are a powerful means of presenting data; first impressions last! But they only give a summary, and this summary can be misleading, either intentionally or by mistake.

Bar charts

The bar chart is one of the most common methods of presenting data in a visual form. A simple bar chart is a chart consisting of a set of non-joining bars. A separate bar for each data item is drawn to a height that is proportional to the frequency of the data item. The widths of each bar are always the same. The bars are usually drawn vertically but they can be drawn horizontally. Figure 7.1 shows the bar chart for the data presented in table 7.1.

Figure 7.1: Bar chart of the gender data presented in table 7.1

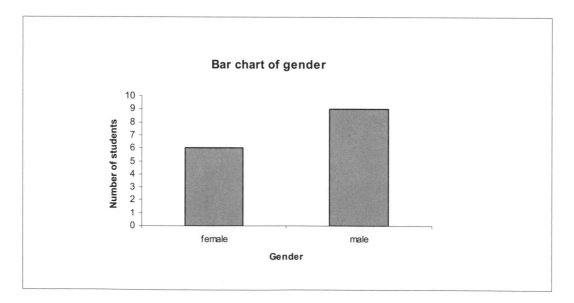

There is little point in constructing bar charts when the variable has only two possible values; the frequency table is actually more informative. However, bar charts are useful for variables with many possible values.

Constructing a bar chart

The following frequency table (table 7.2) shows the nationality of 60 students completing a business studies module at a British university. Draw a bar chart of this data.

Table 7.2: Nationality of students

Nationality	Count	%
American	4	7
British	27	45
Chinese	15	25
Indian	8	13
European	6	10
Total	60	100

Cross-tabulations

More complex tables of counts can summarise data on two variables simultaneously. Such cross-tabulations (or *contingency tables*) allow you to investigate the relationship between the tabulated variables.

The values of one variable define the rows of the table and the values of the other variable define the columns. The number in each *cell* of the table (the intersection of a row and a column) represents the count of the corresponding combination of values. Often row and column totals are included; these give the ordinary frequency distributions of the row and column respectively and are referred to as the *marginal distributions* of the table. Cross-tabulations provide a standard method of summarising the data from a survey and of presenting data in reports and publications.

Example 7.2

Table 7.3 provides a summary of the gender and marital status of the employees in a company. The information relating to the marital status of the employees is in the rows, and the information relating to their gender is in the columns.

Table 7.3: Tabulation of gender against marital status for a company

Marital status	Gender		Total
	Male	Female	
Single	1	1	2
Married	*10*	2	12
Widowed	1	0	1
Total	12	3	15

When you've constructed a cross-tabulation it's important to check that the row totals and the column totals add up to the same grand total. So, in example 7.2 the row totals add up to 2+12+1 = 15, and the column totals also add up to 12+3 = 15.

It is possible to extend cross-tabulations to include discrete data, as well as category data. It would be a relatively easy exercise to produce a cross-tabulation of *gender* against *number of children* for the employees in a company if information were available for each employee relating to the number of children they have. Table 7.4 is an example of such a table and demonstrates that numerical variables can be used in cross-tabulations.

Table 7.4: Tabulation of gender and number of children

Gender	Number of children				Total
	0	1	2	3 or more	
Male	3	4	4	1	12
Female	1	1	0	1	3
Total	4	5	4	2	15

Continuous numerical variables could also be included in cross-tabulations if you categorise them first into groups. For example, you could produce a cross-tabulation showing age and gender of the employees if the numerical variable age is first grouped into intervals such as 20-29, 30-39, 40-49 etc. We look at examples like this at the end of the chapter.

Summarising in percentages

A cross-tabulation can be summarised by calculating percentages of the row or column totals. This method allows two or more groups to be compared.

Example 7.3

Consider again table 7.3 that provided information on the gender and marital status of employees in a company. The table is reproduced below. It is quite difficult to compare the distributions of marital status of females and males because of the different total numbers of females and males. For example, there are 10 married men in this company as opposed to two married women. Is it reasonable to conclude that male employees are more likely to be married? Probably not: there are more men in the company so it is not surprising that there are more married men.

Table 7.5: Tabulation of gender against marital status for a company

Marital status	Gender		Total
	Male	Female	
Single	1	1	2
Married	10	2	12
Widowed	1	0	1
Total	12	3	15

Converting each column to a percentage makes it easier to compare the marital status of males and females. Table 7.6 details the results.

Table 7.6: Percentage comparison of marital status for male and female employees

Marital status	Gender		Total
	Male	Female	
Single	$\frac{1}{12} \times 100 = 8\%$	$\frac{1}{3} \times 100 = 33\%$	$\frac{2}{15} \times 100 = 13\%$
Married	$\frac{10}{12} \times 100 = 83\%$	$\frac{2}{3} \times 100 = 67\%$	$\frac{12}{15} \times 100 = 80\%$
Widowed	$\frac{1}{12} \times 100 = 8\%$	0	$\frac{1}{15} \times 100 = 7\%$
Total	100%	100%	100%

Note: Percentages may not sum to totals due to rounding.

From this table we can see that male employees are more likely to be married than female employees (because their percentage is higher) and female employees are more likely to be single than male employees (similarly).

Activity 7.8

Using percentages with cross-tabulations

Use percentages to compare the distributions of number of children for male and female employees using the information in table 7.4.

Activity 7.9

Laptop survey

The laptop survey in activity 7.5 also asked people what make of laptop they were using. Table 7.7 gives the results as a cross-tabulation of operating system and laptop make.

Table 7.7: Tabulation of operating system and laptop make

Laptop make	Operating system			Total
	Windows	MacOS	Linux	
Dell	5	1	1	7
HP	1	1	1	3
IBM	4	0	0	4
Sony	1	5	0	6
Total	11	7	2	20

Use percentages to compare the distributions of laptop operating system.

7.3 Summarising numerical data

Stem and leaf plots

A stem and leaf display is a statistical technique to present a set of numerical data. Each numerical value is divided into two parts. The leading digit(s) becomes the stem and the trailing digit(s) the leaf. The stems are located along the vertical axis, and the leaf for each observation along the horizontal axis. This idea should be clearer to you by working through the next example.

Example 7.4

Consider the following ages (in years) of 25 part-time employees.

63	27	46	47	22	64	30	19	69	36	65	60	40
66	55	33	47	42	49	23	22	46	62	30	20	

Source: Hypothetical data

The data consists of two-digit numbers, so it is fairly obvious how we will split the numbers into stems and leaves. The first digit (the '*tens*' digit) will form the stems and the second digit (the '*units*' digit) will form the leaves. For example, for the data item 63, the stem is the 6 and the leaf is 3. For a stem and leaf display, write all the possible stems, in order, on the left-hand side of a vertical line. The smallest number in this data set is 19 (stem 1) and the largest is 69 (stem 6) so the possible stems go from 1 through to 6.

Then go through the data values, in the order they are given, and record the leaf of the value opposite the corresponding stem. The first five values (63,27,46,47,22) are put on like this:

```
1  |
2  | 7   2
3  |
4  | 6   7
5  |
6  | 3
```

When all the values have been recorded, the display looks like this:

```
1  | 9
2  | 7   2   3   2   0
3  | 0   6   3   0
4  | 6   7   0   7   2   9   6
5  | 5
6  | 3   4   9   5   0   6   2
```

It is a good idea now to order the leaves on each stem from smallest to largest and include a count column which indicates how many data items are on each stem. Then to finish it all off, include a title as shown in figure 7.2.

Figure 7.2: Stem and leaf display of ages of part-time employees in years

								Count
1	9							(1)
2	0	2	2	3	7			(5)
3	0	0	3	6				(4)
4	0	2	6	6	7	7	9	(7)
5	5							(1)
6	0	2	3	4	5	6	9	(7)

Notice that, as well as sorting the data into order, the stem and leaf provides a visual display of the data: it is easy to compute the numbers of employees in different age groups. (Note that it is therefore essential to space out the leaves evenly.) The leaves on each stem can be counted and these counts have been shown in brackets on the right of the leaves. The counts show the frequency associated with each stem. In this example, the counts show the number of employees in each age group.

So there is one employee in the age range 10-19, five in the age range 20-29 etc. If you take this information and present it in a tabular form, you get the results presented in table 7.8.

Table 7.8: Distribution of ages of 25 part-time employees

Age of employee	Number of employees
10-19	1
20-29	5
30-39	4
40-49	7
50-59	1
60-69	7
Total	25

This sort of table is often referred to as a *grouped frequency distribution* in textbooks. A grouped frequency distribution organises the data items into groups or *classes* of values. It then shows how many data items are within each class, which is referred to as the *class frequency.* So table 7.8 tells us that there are five employees aged between 20 and 29 in this company.

Grouping allows you to see any pattern in the data. However, it is important to realise that grouping results in a loss of information. In table 7.8 we know that there are seven employees whose age is between 40 and 49 but, without access to the original data set, we would not know their exact ages. From the stem and leaf plot we could read off the ages of the employees who are aged between 40 and 49. Looking at the 4th row in figure 7.2, we see this has a stem of 4. Therefore, all the numbers on this row are between 40 and 49. Reading them off gives the values 40, 42, 46, 46, 47, 47 and 49. So stem and leaf plots and grouped frequency tables or distributions both allow you to see the general pattern of the data. Grouped frequency distributions result in a loss of some information, whereas stem and leaf plots still allow you to work back to the actual data and, therefore, do not result in a loss of information. However, stem and leaf plots are impractical for very large data sets as the display would be wider than the page and, in such situations, you need to use grouped frequency tables to examine the general patterns in the data.

In these examples we've used groups of size 10, and we've labelled the stem with tens digit of the group, so for example the group 10 – 19 had the label '1' on the stem, and so on.

For other data sets it may be better to use a different set of groups or different labelling. For example, if we're measuring the height of people in centimetres then (almost) all the heights will begin with 1, so it might be better to label the stem with the first two digits, so that the group 150 – 159 is labelled '15', for example.

In other cases it may be better to use larger or smaller groups to get a sensible number of rows on the stem. For example, we may decide to use groups of size 5. One convention here is to label the first group (say 10-14) with a '1' and the second group (say 15-19) with a '1*'. However, there are many different ways to organise the groups and label the diagram, so you need to check carefully that you understand what a particular diagram means.

Activity 7.10

Compiling a stem and leaf plot

The examination marks obtained by a class of students were as follows:

53	77	46	46	42	63	75	60	61	49	34	63	45
63	32	20	50	53	38	54	87	65	44	59	57	57

1. Form a stem and leaf display of this set of data.

2. Describe the main features of this set of examination marks, bearing in mind that the pass mark was 40% and that marks over 70% were awarded a distinction.

7.4 Histograms

Consider again the stem and leaf display of the employee ages in figure 7.2. Suppose boxes are put around each row of leaves as follows:

Figure 7.3: The stem and leaf display of the employee ages

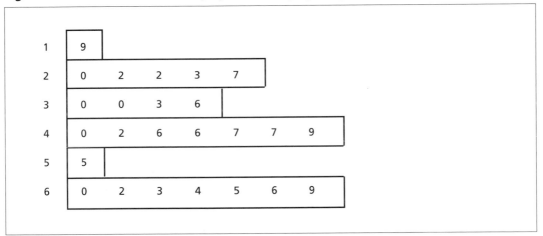

Now remove the leaves but keep the boxes. This shows the shape of the stem and leaf display but not the individual values. Each stem is represented by a box or *bar* whose length represents its frequency. The resulting figure 7.4 is a *histogram* of the data. Notice that the stems have been replaced by intervals, like 20-29, which in this example represent age groups.

Figure 7.4: Histogram of ages of 25 employees

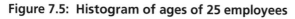

Histograms are more usually drawn with vertical, rather than horizontal, bars, as in figure 7.5.

Figure 7.5: Histogram of ages of 25 employees

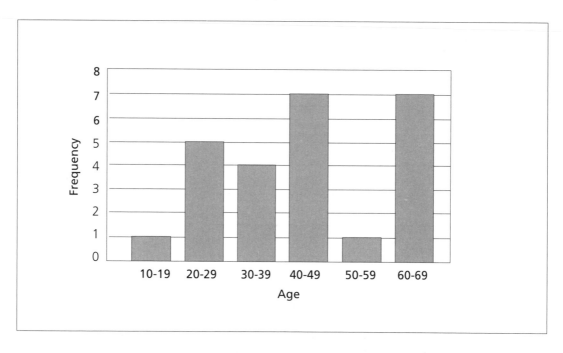

As with a grouped frequency distribution, a histogram conveys less information than the stem and leaf. However, histograms can be used for large data sets where stem and leaf displays would be too wide for the page.

Drawing a histogram

Consider again the exam data you worked on in activity 7.10. Draw a histogram of these examination marks.

Activity 7.11

Activity 7.12

Emails

A group of students were asked how many email messages they had sent in the last month. The results were as follows:

23	12	45	36	3	17	0	17	40	4	29	10
25	5	20	14	0	65	38	8	24	22	12	1

Sort these results into suitable frequency classes and draw a histogram of them.

Example 7.5

A company is considering changing its policy on flexible working times. The bar chart in figure 7.6 shows the number of staff who are in favour of the new proposals, and the number who are against them.

Figure 7.6: Bar chart comparing numbers of staff in favour of and against new policies on flexible working

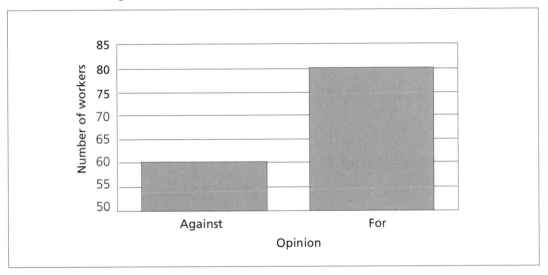

The visual presentation of the graph could lead us to conclude that three times as many staff are in favour of the proposal, as the 'in favour' bar is three times higher than the 'against' bar. However, if we look at how many staff are actually in favour of, or against, the proposals we see that 80 are in favour and 60 are against. In percentage terms

$$\frac{80}{140} \times 100 = 57\%$$ are in favour and

$$\frac{60}{140} \times 100 = 43\%$$ are against, so in fact this is closer to an even split.

The problem with the bar chart is that the scale on the y-axis (frequency) does not start from zero. As a result, the difference in the heights of the two bars has been exaggerated. A fairer way to draw the bar chart is shown in figure 7.7:

Figure 7.7: Fairer bar chart comparing numbers of staff in favour of and against new policies on flexible working

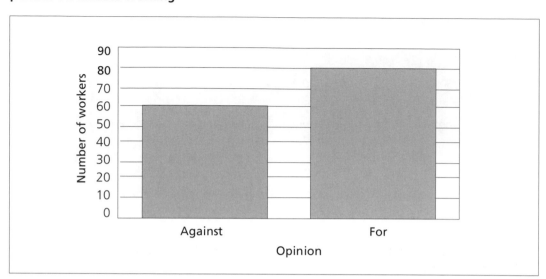

7.5 Summary

In this chapter we looked at different types of data, and saw ways to summarise data to find their essential features. We saw some ways to use graphs and diagrams to represent features of data in helpful ways. These ideas about data will be used throughout the next few chapters.

7.6 Review questions

 Question 7.1: A class of 31 students is given a test that has 20 questions on it. The number of correct answers for each students are recorded. What type is the resulting variable?

    A: A nominal category variable.

    B: An ordinal category variable.

    C: A discrete numerical variable.

    D: A continuous numerical variable.

 Question 7.2: A class of 31 students is given a test that has 20 questions on it. At the end of the test, students are asked if the test was harder than expected, as expected or easier than expected, and their answers recorded. What type is the resulting variable?

    A: A nominal category variable.

    B: An ordinal category variable.

    C: A discrete numerical variable.

    D: A continuous numerical variable.

Question 7.3: A class of 31 students is given a test that has 20 questions on it. The time each student took to complete the test is recorded. What type is the resulting variable?

   A: A nominal category variable.

   B: An ordinal category variable.

   C: A discrete numerical variable.

   D: A continuous numerical variable.

Question 7.4: Assessing different types of data

Consider the following variables. Decide whether the type of data for each one is a category, discrete numerical or continuous numerical. If you decide it is a category variable, explain whether it is nominal or ordinal data.

1. The time athletes recorded for running the final of the 100m in the last Olympic Games.

2. The country the athletes represented.

3. The number of times the athlete had represented their country in the 100m at the Olympic Games.

4. The athlete's impression of how fast they ran: faster than expected, as fast as they expected, or slower than they expected.

Question 7.5: Constructing and reading bar charts

A company has completed a survey of its employees' opinions on their attitudes to their working experience with the company. The company managers know that their employees will be more productive, and likely to stay with the company long term, if staff morale is high and staff have a positive view of what it is like to work for the company. The following frequency distributions provide the following information.

Table 7.9: Responses of employees regarding factors which influence job satisfaction

Important factors in job satisfaction	Frequency
Flexible working	32
Pleasant working environment	37
Higher salary	8
Career development opportunities	14

Table 7.10: Important factors in creating a pleasant working environment

Desirable features of the working environment	Number of employees selecting the feature
Air conditioning	35
Open plan working space	7
Well-equipped, pleasant staff room	21
Refectory with wide choice and value for money	10
No-smoking policy	18

1. Draw a bar chart for each distribution.

2. Write a brief summary to management explaining what actions could be taken to improve staff morale.

Question 7.6: Interpreting cross-tabulations

A mail order company conducted a survey of 600 of its customers about the service it provides. One of the questions asked was if the customers had been experiencing any problems with the service within the last six months. The company tabulated this information, along with whether the customer was a regular or an occasional customer. Table 7.11 shows the partial cross-tabulation.

1. Fill in the remaining cells and marginal totals in table 7.11.

2. What is the marginal distribution of whether the customers surveyed were occasional or regular customers?

3. Use percentages to enable you to compare the distributions of problem rate for occasional and regular customers. If a customer has experienced a problem in the last six months, is it more likely to be an occasional or a regular customer?

Table 7.11: Cross-tabulation of type of customer against whether problems had been experienced with the service

Problem	Type of customer		Total
	Occasional	Regular	
Yes		240	400
No		80	200
Total			

Based on an example in Johnson and Wichern,
Business Statistics, decision making with data, Wiley press.

Question 7.7: Stem and leaf displays

The following figures represent the length of time (in minutes) that a sample of 32 students took to answer an exam question.

14 31 28 29 20 17 21 83 23 12 15 33 75 14 62 35
27 15 18 64 15 21 81 22 23 73 24 28 23 41 26 26

Draw a stem and leaf plot of these times. Students were expected to take at least 20 minutes to answer the question. If they managed to answer the question quicker than that, the tutor would expect them not to get a passing mark for their answer. What percentage of students did the tutor suspect would fail the question?

 Question 7.8: Histograms

A company buys in electrical components in batches of 2,000. From time to time a batch is randomly selected and, for quality control purposes, all the components are inspected. The data below gives the number of defective components found in 40 batches recently bought in from supplier A.

12	16	81	49	60	17	19	48
34	20	25	50	32	72	57	44
76	62	93	43	47	93	86	71
54	66	48	51	27	22	16	53
48	61	33	19	78	49	98	19

1. Draw a histogram of this data.

2. A batch is returned to the supplier if 60 or more defectives are found in the batch. If more than 30% of the batches are returned, the company will contact the supplier to discuss the quality of the goods being supplied. Based on the information from these 40 batches, does the company need to contact the supplier? Explain your answer.

7.7 Answers to review questions

Question 7.1: The answer is C.

Values in the variable will be similar to 10, 12, 18, 20 etc. The variable is created by counting and recording how many answers each student got right. A count variable is discrete.

Question 7.2: The answer is B.

The answer given must be in one of three categories: harder than expected, as expected, or easier than expected. Presumably, for the students, they will be happier if the test is easier than expected rather than harder than expected so there is an order among these categories. Therefore, the variable is an ordinal category variable.

Question 7.3: The answer is D.

This variable is a measurement of the amount of time taken. Therefore, it is a continuous numerical variable.

Question 7.4: Assessing different types of data

1. *Time recorded:* this is a numerical variable. Time is a measurement which could be measured to an endless number of decimal places, if a timing device was available that recorded the times that accurately. Therefore, time is a continuous numerical variable.

2. *Country represented:* this is a category variable, as athletes will fall into only one category such as Britain, Canada, France etc. There is no order between these categories, i.e. representing Canada is not better than representing Britain, so this is a nominal variable.

3. *Number of times the athlete had represented their country:* this is a numerical variable. As it is a count, it is a discrete numerical variable.

4. *Athlete's impression of how fast they ran:* this is again a category variable as answers will be in one and only one group: faster than expected, as fast as they expected, or slower than expected. Clearly, running faster than expected is better than running slower than expected. Therefore, there is an implied order between these categories, so this is an ordinal variable.

Question 7.5: Constructing and reading bar charts

1. The bar charts are produced in figures 7.8 and 7.9.

Figure 7.8: Important factors in job satisfaction

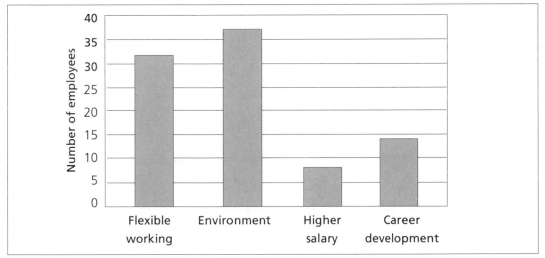

Figure 7.9: Working environment factors

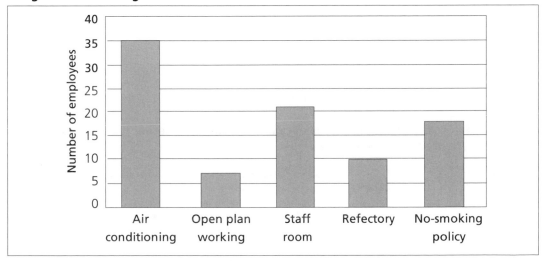

2. From the first bar chart, the most important factors in terms of morale for the staff are a pleasant working environment and flexible working hours. From the second bar chart, the important environmental considerations for the staff are air conditioning, followed by a decent staff room and a no-smoking policy. To improve the environment, and hence staff morale, the company should prioritise investing in air conditioning and the staff room facilities. They also need to look at the most important issues relating to flexible working.

Question 7.6: Interpreting cross-tabulations

1. The completed cross-tabulation is shown in table 7.12. The values in italics have been calculated from the information in table 7.11.

Table 7.12: Cross-tabulation of type of customer against whether problems had been experienced with the service

| Problem | Type of customer | | Total |
	Occasional	Regular	
Yes	160	240	400
No	120	80	200
Total	280	320	600

2. The marginal distribution of whether the customers surveyed were occasional or regular customers is shown in table 7.13.

Table 7.13: Marginal distribution of customer status

Type of customer	Count	%
Occasional	280	47
Regular	320	53
Total	600	100

3. The type of customer is in the columns of table 7.13. Therefore, to compare the problem rate for the different type of customers we need to use the column totals to convert table 7.12 into percentages. The result is shown in table 7.14.

Table 7.14: Cross-tabulation of type of customer against whether problems had been experienced with the service

| Problem | Type of customer | | Total |
	Occasional	Regular	
Yes	57%	75%	67%
No	43%	25%	33%
Total	100%	100%	100%

Regular customers (75%) are more likely to have experienced problems in the last six months than occasional customers (57%).

Question 7.7: Stem and leaf dispalys

Figure 7.10 presents the correct stem and leaf plot for this data.

Figure 7.10: Stem and leaf plot of times taken to answer the question

														Count
1	2	4	4	5	5	5	7	8						(8)
2	0	1	1	2	3	3	3	4	6	6	7	8	8	9	(14)
3	1	3	5											(3)
4	1													(1)
5														(0)
6	2	4												(2)
7	3	5												(2)
8	1	3												(2)
												Total = 32		

From this stem and leaf we can see that eight students completed the question in less than 20 minutes. Therefore, the tutor is expecting 8 out of 32, which is 25%, of the students to fail the question.

Question 7.8: Histograms

1. You may find it easier to compile the stem and leaf display first before drawing the histogram. Figure 7.11 details the stem and leaf display.

Figure 7.11: Stem and leaf display of number of defective items

									Count
1	2	6	6	7	9	9	9		(7)
2	0	2	5	7					(4)
3	2	3	4						(3)
4	3	4	7	8	8	8	9	9	(8)
5	0	1	3	4	7				(5)
6	0	1	2	6					(4)
7	1	2	6	8					(4)
8	1	6							(2)
9	3	3	8						(3)
								Total = 40	

This can then be converted into a histogram by putting boxes around the leaves in each stem as shown in figure 7.12.

Figure 7.12: Histogram of the number of defective items

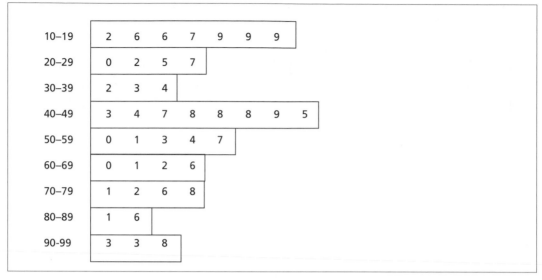

Alternativel,y you can construct a histogram directly and maybe display it vertically, as shown in figure 7.13.

Figure 7.13: Histogram of the number of defective items

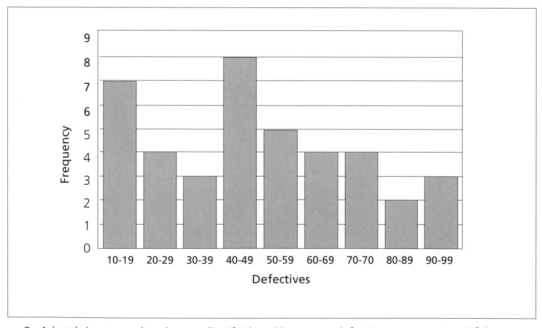

2. A batch is returned to the supplier if it has 60 or more defective components. Of these 40 batches, 13 (4+4+2+3) have 60 or more defective items. As a percentage this is $\frac{13}{40} \times 100 = 32.5\%$, which is greater than 30%, therefore the supplier needs to be contacted.

7.8 Feedback on activities

Activity 7.1: Data relating to you

Few of your answers were given in the form of numbers. Probably the only numerical answers you gave were for parts 1 and 5, and part 6 which required a list of numbers (hopefully, you've completed more than one of the activities by now!). However, all your answers are precise data relating to you, which might be of interest to an employer, a market researcher etc.

Activity 7.2: Identifying types of data

1. This is a numerical variable. As it is a count, it is a discrete variable.
2. This is a category variable as people will fall into only one category such as blue, brown, green etc. There is no order between these categories, i.e. green eyes are not better than blue eyes so this is a nominal variable.
3. This is again a category variable as answers will be in one and only one group, i.e. strongly disagree, disagree, don't know, agree, strongly agree. Depending on the purpose of the question and how you will report the answers, you will able to decide if someone agreeing is better than disagreeing (or vice versa). Therefore, there is an implied order between these categories: agree is better than disagree, so this is an ordinal variable.
4. This is a numerical variable. As it is a measurement, it is a continuous variable.
5. This is a numerical variable. As it is a count, it is strictly a discrete variable. However, as the counts are so large it will probably be treated as a continuous variable in any analysis.

Activity 7.4: Creating a frequency table

You should have a frequency table which looks like table 7.15.

Table 7.15: Gender of students in a class

Gender	Count	%
Female	6	$\frac{6}{30} \times 100 = 20\%$
Male	24	$\frac{24}{30} \times 100 = 80\%$
Total	30	100%

Comparing the number of female students in table 7.15 with those in table 7.1 we can see that there are six female students in each class. So, based on counts, there do appear to be the same number of females in each class. However, this comparison of the actual number of females in each class ignores the fact that the classes are actually of different sizes. The second class is twice as big as the first, 30 students compared to 15. This difference in the totals is reflected in the percentages with 20% of the second class being females in comparison to 40% of the first class being female. Therefore, the balance of male/female students in each class is more apparent if the percentages are used for the comparison rather than the raw totals.

Activity 7.5: Creating a frequency table 2

You should have a frequency table which looks like table 7.16.

Table 7.16: Choice of operating system

Operating system	Count	%
Windows	11	$\frac{11}{20} \times 100 = 55\%$
MacOS	7	$\frac{7}{20} \times 100 = 35\%$
Linux	2	$\frac{2}{20} \times 100 = 10\%$
Total	20	100%

Windows is the most popular operating system, since it is used by the largest number of people (11). Linux is the least popular operating system, since it is used by the smallest number of people (2). It's probably just about possible to answer these questions from the original data, but it's much clearer and more reliable to use the frequency table, particularly when there are many different items in the data.

Activity 7.6: Constructing a bar chart

Your bar chart should look similar to the one presented in figure 7.14.

Figure 7.14: Nationality of a class of students

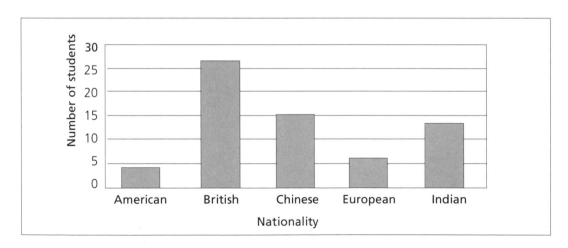

Activity 7.7: Reading a cross-tabulation

1. The value '10' is in the cell corresponding to the intersection of the male gender column and the marital status married. Therefore, there are 10 married men in this company.

2. You could answer this question by adding up all the values in the cells that correspond to the genders and marital status intersecting, i.e. 1+1+10+2+1+0 = 15. Alternatively, the value in the bottom right-hand cell relating to the totals gives you the overall total for the data, which is 15.

3. For this marginal distribution we are only interested in how many employees there are for each marital status. As marital status is in the rows, this means we are only interested in how many people there are on each row of the table. The totals for the rows is given in the total column on the right-hand side of the table. Table 7.17, therefore, gives the marginal frequency table for marital status.

Table 7.17: Frequency table for marital status

Marital status	Count
Single	2
Married	12
Widowed	1
Total	15

A percentage column could easily be added to this table. Likewise, a similar marginal frequency table could be produced for gender.

Activity 7.8: Using percentages with cross-tabulations

Table 7.4 (reproduced here) shows gender in the rows of this table, therefore to compare genders we need to take percentages using the row totals as shown in table 7.18.

Table 7.4: Tabulation of gender and number of children

Gender	Number of children 0	1	2	3 or more	Total
Male	3	4	4	1	12
Female	1	1	0	1	3
Total	4	5	4	2	15

Table 7.18: Percentage comparison of number of children for male and female employees

Gender	Number of children 0	1	2	3 or more	Total
Male	$\frac{3}{12} \times 100$ = 25%	$\frac{4}{12} \times 100$ = 33%	$\frac{4}{12} \times 100$ = 33%	$\frac{1}{12} \times 100$ = 8%	100%
Female	$\frac{1}{3} \times 100$ = 33%	$\frac{1}{3} \times 100$ = 33%	$\frac{0}{3} \times 100$ = 0%	$\frac{1}{3} \times 100$ = 33%	100%
Total	$\frac{4}{15} \times 100$ = 27%	$\frac{5}{15} \times 100$ = 33%	$\frac{4}{15} \times 100$ = 27%	$\frac{2}{15} \times 100$ = 13%	

Note: Percentages may not sum to totals due to rounding.

Activity 7.9: Laptop survey

Table 7.7 is reproduced, showing operating system in the columns of this table; so to compare operating systems we need to take percentages using the column totals as shown in table 7.19.

Table 7.7: Tabulation of operating system and laptop make

| Laptop make | Operating system | | | Total |
	Windows	MacOS	Linux	
Dell	5	1	1	7
HP	1	1	1	3
IBM	4	0	0	4
Sony	1	5	0	6
Total	11	7	2	20

Table 7.19: Percentage comparison of laptop make given choice of operating system

| Laptop make | Operating system | | | Total |
	Windows	MacOS	Linux	
Dell	$\frac{5}{11} \times 100$ $=45\%$	$\frac{1}{7} \times 100$ $=14\%$	$\frac{1}{2} \times 100$ $=50\%$	$\frac{7}{20} \times 100$ $=35\%$
HP	$\frac{1}{11} \times 100$ $=9\%$	$\frac{1}{7} \times 100$ $=14\%$	$\frac{1}{2} \times 100$ $=50\%$	$\frac{3}{20} \times 100$ $=15\%$
IBM	$\frac{4}{11} \times 100$ $=36\%$	$\frac{0}{7} \times 100$ $=0\%$	$\frac{0}{2} \times 100$ $=0\%$	$\frac{4}{20} \times 100$ $=20\%$
Sony	$\frac{1}{11} \times 100$ $=9\%$	$\frac{5}{7} \times 100$ $=71\%$	$\frac{0}{2} \times 100$ $=0\%$	$\frac{6}{20} \times 100$ $=30\%$
Total	100%	100%	100%	100%

Activity 7.10: Compiling a stem and leaf plot

1. Figure 7.15 shows what the stem and leaf display should look like for this data.

2. In total, 22 (84.6%) of these students passed the exam. (If the pass mark is 40%, 6+7+6+2+1 = 22 passed.)   Distinctions (marks above 70) were awarded to three (12%) students.

Figure 7.15: Stem and leaf display of examination marks out of 100%

								Count
2	0							(1)
3	2	4	8					(3)
4	2	4	5	6	6	9		(6)
5	0	3	3	4	7	7	9	(7)
6	0	1	3	3	3	5		(6)
7	5	7						(2)
8	7							(1)

Activity 7.11: Drawing a histogram

You should have drawn something similar to the histogram presented in figure 7.16. This histogram is approximately *symmetric* and *bell-shaped* – it has a single peak in the middle with the frequencies trailing off each side. This means that this distribution can be approximated by a normal distribution – an important theoretical distribution defined by a mathematical formula.

Figure 7.16: Histogram of examination marks

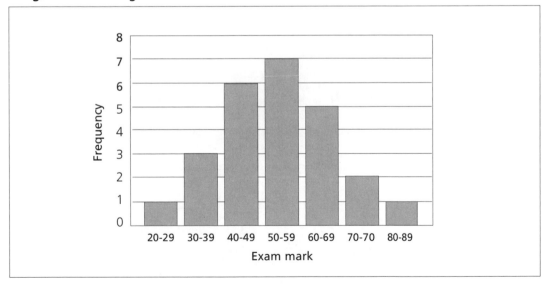

Activity 7.12: Emails

The are various ways we could group the data, but the obvious one is to use groups of 10 from 0..9 then 10..19, and so on. We get table 7.20:

Table 7.20: Group frequency table for email data

Number of emails	Frequency
0 - 9	7
10 - 19	6
20 - 29	6
30 - 39	2
40 - 49	2
50 - 59	0
60 - 69	1

Notice that noone sent between 50 and 59 emails. This is much easier to see in the frequency table than it was in the original data. The histogram is shown in figure 7.17.

Figure 7.17: Histogram of number of emails sent

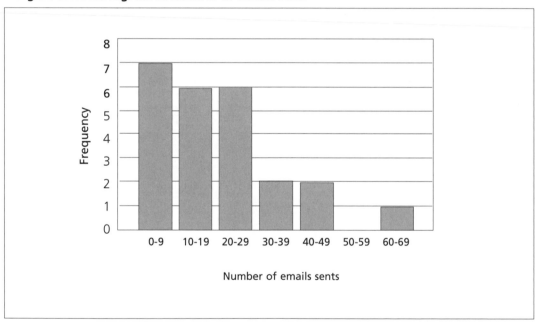

CHAPTER 8

Summary statistics 1 – measures of location

OVERVIEW

In the last chapter we considered different types of data and looked at various ways of presenting them.

This chapter will also use these data types, but we will attempt to summarise them by using just a few numbers to represent the data, rather than tables or graphs.

We will see that, while useful, these summary statistics are frequently inadequate and this will lead us to consider further types of summary statistics in chapter 9.

Learning outcomes	On completion of this chapter, you should be able to:

- Interpret the measures of location for a set of data

- Find the value of the mean, median and mode for a given data set

- Describe what each of these tells us about the data

- Recognise the limitations of each of these measures

- Choose the most appropriate method of location for a particular task.

8.1 Measures of location

In statistics we use sets of data all the time. Some sets are quite small and easy to deal with. We can look at a few numbers, and without doing too much work, we can often make sense of what's going on. But some sets of data are quite big. In fact, some are absolutely huge. It would be impossible to look at a set of a few thousand (or maybe just a few hundred) numbers and get a good idea of what's going on.

Now the whole point of statistics is that it's supposed to help us make sense of all this data. One of the things it can do rather well is give us ways of presenting data so that we can make sense of large sets of numbers.

You've already seen a few methods of summarising data in chapter 7. These methods include putting the data into some kind of order or pattern, constructing tables or drawing pictures that represent some aspect of the data. You've also seen some of the ways these methods can be misused, if not actually to lie about a situation, then, at least, to make it difficult to see the truth.

The methods used in the last chapter are fine if you're writing a report or giving a presentation. But, they do have their limitations.

Suppose someone wanted to know something about the ages of the people in your class, and you just happen to have collected the data. Well, you could whip out a carefully prepared bar chart you just happened to have on you. Or, and this would make you seem a lot less strange, you could simply give the answer as some kind of number.

In fact, we do this all the time. We talk about what things are most likely to happen, or what the average value is, and all the time we are relying on some numerical summary of a set of data.

What we would really like is a way to summarise a whole set of data without having to draw a lot of pictures or graphs, and do it in a way that helps people to understand what is going on, rather than confusing them, or even lying to them.

And that's what this chapter is about.

We are going to look at three different ways of summarising a set of data by just using one number.

That doesn't mean that we have three different ways of finding the same number. What it means is that we can find three *different* numbers, all of which can tell us something *different* about the data. Sometimes those numbers will be the same as each other, but often they will be different, and that difference can also tell us useful things about the data.

These numbers are called **measures of location**. Some people call them *measures of central tendency*, but it means the same thing.

The measures of location we are going to look at are the:

- **Mode**
- **Median**
- **Mean**.

8.2 Looking forward to chapter 9

You might be wondering why, if these numbers can tell us so much, did we bother with all those methods of summarising in the last chapter.

Well the numbers we are going to look at can tell us a lot, but they don't tell us everything. Sometimes we need to know more than just a measure of location; we also need to know about how spread out the data is. We will be doing this with more numbers in the next chapter. We will also be using some graphical methods to make all these measures easier to understand.

What's more, we will actually be using some of the things you did in the last chapter to help us find some measures of location and spread. You will often need to go back to the work you did in the last chapter, and refer to the notes you wrote in your *learning journal*. We will be using them a lot.

You really need to think of this chapter as the middle section of a three-chapter set.

8.3 Before you start...

In this chapter you should collect some data from your friends or from fellow students on your course. You'll want to use this data to try out some of the methods in this chapter, and also the next chapter on measures of dispersion. You could just get the data for a few friends or family members. However, if you know other people who are studying from this book then you could share your data with them, so that you all have more data to work with.

You need to find out the following about each person:

- **What is their age?**

  Of course, you may not want to give your real age, and there's nothing to stop you lying about it. But the work you are going to do will be much more interesting if you have a variety of responses.

- **How many siblings (brothers and sisters) do they have?**

  It's up to you whether you want to include half-brothers or step-sisters, adopted or foster children. Remember that the only reason you are collecting the data is so that you can work with it.

- **What is their height?**

  You must decide what measurements to use. It doesn't really matter whether you use centimetres or feet and inches, as long as all the data uses the same measurements. If you're sharing data with someone else, then you need to agree to use the same units. We don't want anyone who is 6cm or 120 feet!

- **How far do they live from their workplace or study centre?**

  Again, you must decide whether to work in miles or kilometres, and be consistent.

- **How many sets of observations are there?**

  You will need to know how many items of data you have for each of these questions. You should have the same number for all of them, but occasionally a piece of data might be missing for a particular person, and this will affect the counts.

8.4 The mode

The mode is simply the data result that occurs most often. You may not realise it, but in a sense, it's something you use all the time.

The modal response for categorical data

Go back to activity 7.6 in chapter 7 – *Summarising and presenting data*. In that activity you were given the following frequency table:

Table 7.2: Nationality of students

Nationality	Count	%
American	4	7
British	27	45
Chinese	15	25
Indian	8	13
European	6	10
Total	60	100

You were asked to produce a bar chart which should have looked something like this:

Figure 7.14: Nationality of a class of students

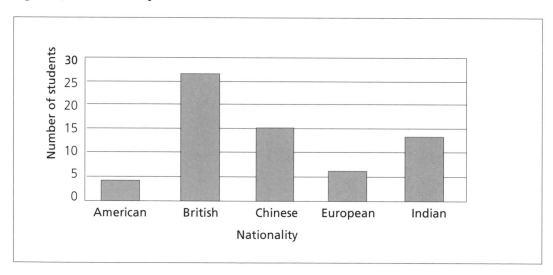

Now, if we asked you what nationality formed the biggest group, you should be able to tell us that it was the British. The British students accounted for 27 of the 60 students, with a relative frequency of 45%. Alternatively, you could quite simply say that the biggest bar in the bar chart represented British students.

Now, we're supposed to be looking for numbers, and the word 'British' is hardly a number. So it can't quite be the mode for this data. But, it is the *modal response*, or the most common thing.

Category data can't really have a measure of location in the same way as numeric data, but the way we identify the mode for numeric data is exactly the same as the way we just identified the biggest group in the category data.

The mode for numeric data

Suppose we were writing material for students working on a foundation course. We would like to know what sort of ages we were writing for so that we could use language and examples that are relevant to that age group.

There are several ways we could ask about the ages of the group, and each of these questions could have a slightly different meaning:

- What is the average age of the students?
- How old is the student whose age is in the middle?
- How old are most of the students?
- If we had to guess the age of an unseen student chosen at random, what would be the most sensible guess?

In those questions, we have already hinted at what some of the other measures of location can tell us, but the one we are interested in here is the last one. Our best guess, the one that's most likely to be right, is the value that occurs most often.

Activity 8.1

Finding the mode of a set of data

Here is a list of ages of students in a class:

Data set 8.1: Ages of a hypothetical set of students

18	22	18	19	25	20	19	18	20	22
21	19	18	20	20	36	19	21	20	18
19	18	18	21	23	18	21	18	20	19

1. What is the age that occurs most often?

2. Can you say how old most of the students are, using a single number?

Activity 8.2

Finding the mode 2

Here is the number of emails sent in a week by the workers in a company:

Data set 8.2: Number of emails sent in a hypothetical company

1	12	22	8	0	16	24	4
14	3	1	11	13	29	3	9
10	21	19	2	15	1	16	7

1. What is the number that occurs most often?

2. Does this give a good idea of the number of emails sent by each worker?

The mode for grouped data

We have now seen how to find the mode for a set of data, but we have also seen that sometimes the mode doesn't appear to be very important.

To tell us something significant, we really want to be dealing with just a few values, and we want the frequency of the mode to be quite a lot higher than most of the other frequencies.

Let's go back to the exam marks data in chapter 7 in figure 7.15.

Figure 7.15: Stem and leaf display of examination marks out of 100%

								Count
2	0							(1)
3	2	4	8					(3)
4	2	4	5	6	6	9		(6)
5	0	3	3	4	7	7	9	(7)
6	0	1	3	3	3	5		(6)
7	5	7						(2)
8	7							(1)

Can you see a modal value for this data? Three students score 63%. But two students scored 46%, two more got 53% and another two got 57%. None of this seems to be telling us anything important about what the data looks like, because we have such a wide range of possible values, and only 26 students.

Very often, with data of this kind, we may not see a mode at all. It's quite possible that every student will have a different mark. So does that mean we can't use the mode at all?

Think back to the modal response for the nationality data. Although we didn't have numbers, we could find a kind of mode by looking at the group with the highest frequency, or the biggest bar in the bar chart. We can do the same with numeric data, if it is grouped. We can simply ask ourselves, 'Which is the biggest group?'

In this case, the biggest group is the 50–59% group, so this is the *modal group* or *modal class*.

It may seem like this isn't very important, since its frequency is only one more than the groups on either side. However, we will see later that this combination contributes to the shape the data takes, and that can be very important.

Finding the modal class for a set of grouped data

A class of 32 students took a multiple choice maths test with 50 questions. Here are the number of questions each student got right.

Data set 8.3A: Multiple choice test marks (set A) out of 50

2	37	41	2	18	2	14	7
2	36	11	43	26	37	17	27
24	17	44	20	18	6	35	13
17	18	47	21	0	10	45	11

Using either a frequency distribution chart or a stem and leaf plot, find the modal group for this data.

You should try this using two different methods for grouping the data.

Problems with the mode

We have already seen some of the limitations of the mode:

- The mode tells us which is the most common value, but that doesn't mean that most of the data has that value
- If we have a large range of values, and relatively small frequencies, the mode may not tell us anything very useful
- Data with a large range of values (including data for continuous variables) needs to be grouped, but…
- The modal class will depend on how we group the data.

There is one more thing we should say about the mode before we move on.

Here is some data collected by a bank. It counted how many times its automatic cash dispenser was used over a period of 30 days:

83	64	84	76	84	54	75	59	70	61
63	80	84	73	68	52	65	90	52	77
95	36	78	61	59	84	95	47	87	60

The frequency distribution looks like this.

Table 8.1: Frequency distribution of ATM use

Times used	No. of days (frequency)
30–39	1
40–49	1
50–59	5
60–69	7
70–79	6
80–89	7
90 or more	3
Total	30

Suppose you wanted to know the number of times the ATM is likely to be used on any day. Your best guess would be the modal class – that would give you the greatest chance of making an accurate prediction.

But here, the modal frequency is 7, and there are two classes with that frequency, 60–69 and 80–89. When this happens, the data is called *bimodal*. There are all sorts of reasons why this happens, but we're not going to think about why for the moment.

The problem is, we don't know which of the classes to use. If they were adjacent to each other, that would be easy. We could simply say the most likely number of times is the two classes together. But they are separated by another, lower-frequency class: the 70-79 class.

In this case, the class separating them only has a slightly lower frequency, so the best answer would be that the machine will probably be used between 60 and 89 times. This is a very wide interval, but with this set of data, it's the best we can do.

8.5 The median

One way of summarising a set of data might be to say what value is right in the middle of it. But what do we mean by 'middle'?

Suppose we ask people in a room how long they've been in their present job. These are their answers, to the nearest year, in the order they gave them.

| 2 | 7 | 5 | 4 | 1 |

Here the middle value is 5. However, suppose we had asked them in a different order. We might have got the same answers, but written down like this.

| 7 | 5 | 1 | 2 | 4 |

Now the middle answer is 1.

This is a problem, because we should just be able to work with the data we've got. Our results should not be affected by when, or in what order, we got it. The only way to make sure that happens when we want to find the middle value is to put the data in order first.

This is called *ranking* or *sorting* the data. Here is the above data set after it has been ranked.

| 1 | 2 | 4 | 5 | 7 |

Now we can find the middle value, or median. It's 4.

Ranking data

The median isn't the only statistic we need to rank data for. There are a whole set of others, and we're going to use some of them later on this course. You, therefore, need to make sure you know how to rank a set of data.

Some people seem to find it easy to put a list of numbers into order – some can even do it in their head. That's fine for small sets of numbers, but for larger sets you should use some systematic method, and one of the most useful is something you've already used – the stem and leaf plot.

Activity 8.4

Ranking a set of data

Here is the list of ages of students you used in activty 8.1:

Data set 8.1: Ages of a hypothetical set of students

18	22	18	19	25	20	19	18	20	22
21	19	18	20	20	36	19	21	20	18
19	18	18	21	23	18	21	18	20	19

Construct a stem and leaf plot of this data so that the data is ranked.

Since we aren't using the stem and leaf plot to find the frequencies, it doesn't matter how you choose to group the data. But, you will find the count useful, so remember to include it.

Finding the median

It's easy to find the median if you've just got a few numbers. You can write them down in a row and count from the ends into the middle. However, if you've got a lot of data, you need some easier way of finding the middle value. Fortunately, we have a formula for this.

You need to know how many values you have in your data set. We call this number n. The median position is found from the formula:

$$\frac{n+1}{2}$$

So, if you have the following set of ranked data

5	12	14	14	15	17	19	22	25

then $n = 9$. So the position of the median is:

$$\frac{n+1}{2} = \frac{9+1}{2} = \frac{10}{2} = 5$$

The median can be found in the 5th position of the ranked data – which means the median equals 15.

That's pretty easy, as long as you have an odd number for n. If you think about it, if you have an odd number of something, there's always a middle one. But what about when n is even?

Go back to our student ages data. We have 30 students, an even number. So:

$$\frac{n+1}{2} = \frac{30+1}{2} = \frac{31}{2} = 15.5$$

There isn't a 'fifteen-and-a-halfth' number, so we'll have to look at the number that lies between the 15th and 16th values. This is where the stem and leaf comes in handy.

From the count, we know that there are 15 numbers in the 1* (or 15–19) row. So the 15th number is the last of these – 19. That means the 16th number is the first one on the next row, which is 20.

So the median lies halfway between 19 and 20. The median equals 19.5.

Activity 8.5

Finding the median for a set of data

Use the stem and leaf plot you did for activity 8.3 to help you find the median of the multiple choice marks data.

Activity 8.6

Finding the median 2

Using the data from activity 8.2, find the median number of emails sent in the company.

What do we need besides the median?

In activity 8.5 you saw that the median could indicate things about the data that the mode didn't. Both the mode and the median are good measures of location, but neither of them is perfect. There are still aspects of the data that they don't tell us about. We need more.

There are other summary statistics that are normally associated with the median. These involve looking at measures of dispersion, or the way the data is spread out. These are dealt with in the next chapter.

8.6 The mean

Although we've left this one till last, it's actually the measure of location you're probably most familiar with. The mean is what you're usually talking about when you ask about the average.

The idea behind the mean is that we take all the data values, and share them out equally between all the items. For example, five people have different amounts of money in their pocket. To the nearest dollar they have:

$5 $7 $10 $14 $19

Suppose we decided to share out the money so they all had the same amount. We could add it all up (the total comes to $55), then share it out by dividing by 5 (the number of people). Everyone would, therefore, have $11.

Although we don't actually share the money out, the mean tells us what everyone would get if we did. Here, the mean amount of money is $11. But, did anyone actually have $11 in his or her pocket? Obviously not. So, although the mean gives the traditional average, it may not represent a real amount. It may not *mean* anything…

Calculating the mean

What we did just now was fairly simple, and there's not much more to calculating a mean than that. But we will have to put it in more formal language if we want to use this method more generally.

When we have a set of data that we've collected, we usually refer to the set with a letter. Because the data can be considered a variable, we use the letters traditionally associated with variables: x, y, z etc. So if we call the set of amounts of money x, then all the individual amounts are x values.

We also need to know how many of these x values there are, and just as we did when we wanted to calculate the median, we'll call this number n.

So to get the mean, we add up all the x values and divide by n. In mathematical language, that looks like this:

$$\frac{\sum x}{n}$$

Don't worry about the Σ. It's the capital letter Sigma, from the Greek alphabet. We use it to mean 'the sum of', or more simply, 'add them all up'.

We also have a special term for the mean when we calculate it this way. If we've called the set x, then the mean is called x bar, and is written like this: $\bar{X}$

So the mean is the sum of all the x values divided by how many there are, or:

$$\bar{X} = \frac{\sum x}{n}$$

Activity 8.7

Finding the mean of a set of data

Let's go back to the student ages data:

Data set 8.1: Ages of a hypothetical set of students

18	22	18	19	25	20	19	18	20	22
21	19	18	20	20	36	19	21	20	18
19	18	18	21	23	18	21	18	20	19

Table 8.2: Frequency distribution of ages of students on a foundation course

Age of student	No. of students (frequency)
18	9
19	6
20	6
21	4
22	2
23	1
24	0
25	1
26+	1
Total	30

cont...

1. Use the information in table 8.2 to find the mean of this data.

2. Compare the mean to the mode and median you found earlier.

3. Does the mean tell you anything about the actual ages of the students? Think about this in comparison with the median and mode.

4. Find the mean, median and mode for the exam mark data you used in activity 7.10 of chapter 7.

5. Draw a bar chart of the age data and compare this with the shape of the histogram you drew in activity 7.11 of chapter 7.

Activity 8.8

Finding the mean 2

1. Using the data from activity 8.2 find the mean number of emails sent in the company.

2. Compare the mode (activity 8.2), median (activity 8.6) and mean for this data. Are they the same? If not, what might this tell you about the data? Do you see any similarity with either set of data in activity 8.7?

Comparing means

In activity 8.7 you saw that, even though the mean doesn't always make sense, because it's a number that doesn't actually appear in the data, sometimes it can be useful when we want to compare data sets.

For example, here is some data from a factory:

A machine is used to make electrical components, which are then put into boxes of 20.

At the start of the day, 30 boxes are selected at random, and the components are tested to see how many are faulty. Here are the results:

Table 8.3: First set of results of faults from random selection of components

Number of defective components	Number of boxes
0	9
1	11
2	5
3	3
4	2

At the end of the day, another 30 boxes are selected to see if the machine is performing as well. This time, we get the following results.

Table 8.4: Second set of results of faults from random selection of components

Number of defective components	Number of boxes
0	6
1	10
2	5
3	5
4	2
5	2

Source: Minett-Smith

Activity 8.9

Using the mean to differentiate between two data sets

Use the previous two frequency distributions to find the mean, median and mode for the number of defective components at the start and end of the day.

Do the values make sense?

Do any of the values tell you something useful?

8.7 Summary

In this chapter we looked at ways to be more specific about the overall features of a data set, such as where the data items are located (measures of location) and how spread out they are (measures of dispersion). We'll go on to look at these ideas in more detail in the next chapter.

8.8 Review questions

Question 8.1: Finding the mode for a set of data

The boss of the accounts department of a large company was concerned that people in the office were taking too long for their lunch break, so she timed everyone, every day for a week.

Everybody in the office is entitled to a one-hour lunch break, but these are the times she noted for the 20 workers (in minutes).

Look at the information in data set 8.4.

Data set 8.4: Time taken for lunch breaks (in minutes)

Monday:

47	54	58	58	58	59	60	61	62	62
62	62	63	64	64	65	65	76	82	87

Tuesday:

50	52	54	56	58	59	59	60	61	61
62	63	65	65	65	66	68	74	80	80

Wednesday:

51	53	58	58	59	59	60	61	62	62
62	63	64	65	70	72	73	80	81	81

Thursday:

48	52	55	58	60	60	62	62	63	64
64	66	67	67	67	69	70	72	77	78

Friday:

15	61	62	63	65	66	66	66	69	71
72	72	74	79	84	85	86	86	95	107

The data has already been sorted into order to help you.

1. Find the modal time taken for lunch break for each day of the week.

2. Does it seem like people are taking too long for their break? Does the mode tell you this?

3. Does the data suggest that the day of the week affects the length of time people take?

4. What is the mode for the entire week? Is it what you expected?

 Question 8.2: Finding the modal class for a set of data

Here are the test scores for another class of students taking the same test as in activity 8.3.

Data set 8.3B: Hypothetical multiple choice test marks (set B) out of 50

3	28	3	3	50	1	42	1
14	5	48	12	33	21	43	50
24	50	3	39	19	44	43	21
47	21	49	49	47	28	26	15

Find the modal score for this class. Try using groups of 5 and 10. Are the modal classes using the two different methods the same?

N.B. Before you start working out the groups, look at the values within the data. Should you use the same groups you used in activity 8.3?

Question 8.3: Finding the median of a set of data

1. In question 8.1 you were given data for lunch break times in data set 8.4. It's easy to find the median for the individual days because of the way they have been presented: already ranked.

   Make a note of theses medians.

   Is there anything that suggests that the day of the week affects how long people take for their lunch break?

   Now rank the complete data set (if you haven't already done so) and find the median of the complete data set.

2. In review question 8.2 you used data set 8.3B to find the modal class of a second set of students, set B, taking a multiple choice test. Use the stem and leaf plot from that exercise to help you find the median for this data.

   Compare this with the median for set A who took the test in activity 8.3.

   Do the medians suggest any differences between how well the two sets did in the test?

   Did the mode or modal class suggest the same thing?

Question 8.4: Finding the mean of a set of data

Go back to the lunch break times data in review question 8.1, data set 8.4.

Find the means for each day, and for the whole week, and add these to the median and mode in a table. (You should be able to use the individual means to work out the weekly mean.)

Compare the means and medians. What do they suggest about the length of time people took for their lunch breaks?

Question 8.5: Comparing the means of two sets of data

Find the means for the two sets of students who took the multiple choice test.

Add these to the median and mode.

What do the three measures of location tell you about the two classes?

Question 8.6: Extreme values

Sometimes, most of the data are values that are very close together. There may be just one or two that are very different from the rest. These are called *extreme values*.

What effect do you think they might have on the measures of location?

Try changing one value in some of the data sets you've used so that it's very different from the rest of the values. What effect does it have on the mean, median and mode?

Question 8.7: Which measure of location should you use?

Here's a set of data from a survey:

0	10	11	0	0	0	2	12	0	0
12	1	0	0	0	0	12	0	11	0

Source: adapted from Jarret

Twenty people were asked how often they had bought a copy of *Statistics Monthly* in the last year.

Calculate the mean, median and mode for this data.

How useful are they for explaining the data?

If you were the magazine publisher, what would you be focusing on?

Question 8.8: How useful is the mean?

People used to talk about the 'average' family in the UK as having 2.4 children. Comment on whether this makes sense. How can you explain what it really means?

Nowadays, the average family in the UK is said to have 1.8 children. What does this tell us about the size of families in the UK over the last few years?

Question 8.9: Finding measures of location for your own data

In section 8.3 you were asked to collect data about yourself and friends, family members or fellow students.

You should now have data for your ages, heights, distance lived from your workplace or study centre and how many brothers and sisters you have.

You are now going to begin to prepare a report on these sets of data.

Begin by finding the measures of location for each set of data. You may need to group some of the data, so make sure you explain how you decided on groupings.

Produce stem and leaf plots, and where appropriate bar charts or histograms. Try marking the positions of the mean, median and mode, and comment on the shape of the data. Is it symmetric, or does the data look like it's been dragged away from that shape? Can you spot anything in the data that might cause this to happen?

8.9 **Answers to review questions**

Question 8.1: **Finding the mode for a set of data**

1. You should have found the following modal values:

Table 8.5: Modal time taken for lunch breaks

Day	Mode (in minutes)	Frequency
Monday	62	4
Tuesday	65	3
Wednesday	62	3
Thursday	67	3
Friday	66	3

2. In every case, the modal value was greater than 60. But look at the frequencies for the modes. They aren't very high. On Monday, four people took 62 minutes for lunch, but three people took 58 minutes.

   Although we have identified the mode, it doesn't seem to be telling us anything very important. There are other things about the data that the mode isn't really covering.

   It does seem like people are generally taking longer for their break than they are allowed. We can tell that because, on every day, more than half the people took longer than 60 minutes.

3. The modal values are all very close together, so at the moment there is nothing to suggest that the day of the week affects how long people take for their lunch breaks.

4. The mode for the whole week is 62. You may have expected it to be higher, but this is one of the limitations of the mode. Sometimes the modal value can be quite different from the rest of the data.

Question 8.2: Finding the modal class for a set of data

Using groups of ten, you should have got the following results.

Table 8.6: Frequency distribution of multiple choice marks (set B)

Mark	No. of students (frequency)	%
1–10	7	22
11–20	4	13
21–30	7	22
31–40	2	6
41–50	12	38
Total	32	100

Note: Percentages do not sum to total due to rounding.

Figure 8.1: Stem and leaf plot of multiple choice marks (set B)

										Count
0	1	1	3	3	3	3	5			(7)
1	2	4	5	9						(4)
2	1	1	1	4	6	8	8			(7)
3	3	9								(2)
4	2	3	3	4	7	7	8	9	9	(9)
5	0	0	0							(3)

For the frequency distribution chart it seemed to make sense to use groupings that meant we didn't have to put scores of 50 on their own. But we couldn't do that with the stem and leaf plot. As a result, the two methods give us slightly different answers.

According to the frequency distribution, the modal score for these students is from 41 to 50. However, according to the stem and leaf plot, it's 40 to 49.

If you used groups of 5, then your results should look like this:

Table 8.7: Frequency distribution of multiple choice marks (set B)

Mark	No. of students (frequency)	%
1–5	6	19
6–10	1	3
11–15	3	9
16–20	1	3
21–25	4	13
26–30	3	9
31–35	1	3
36–40	1	3
41–45	4	13
46–50	8	25
Total	32	100

Figure 8.2: Stem and leaf plot of multiple choice marks (set B)

Stem							Count
0	1	1	3	3	3	3	(6)
0*	5						(1)
1	2	4					(2)
1*	5	9					(2)
2	1	1	1	4			(4)
2*	6	8	8				(3)
3	3						(1)
3*	9						(1)
4	2	3	3	4			(4)
4*	7	7	8	9	9		(5)
5	0	0	0				(3)

This time we get totally different answers. According to the frequency distribution chart, the largest class is for scores between 46 and 50, while the stem and leaf tells us it's scores from zero to four that are most likely.

Which one should you use? That's really up to you. This is one of the problems of working with grouped data. You just have to decide which method is the most appropriate for your data – and then stick to it!

Question 8.3: Finding the median of a set of data

1. Here are the medians for the data. We've included the modes in the table so you can compare them more easily.

Table 8.8: Measures of location for lunch break time data

Day	Median	Mode
Monday	62	62
Tuesday	61.5	65
Wednesday	62	62
Thursday	64	67
Friday	71.5	66
Whole week	63.5	62

We've already seen that the modal value may not be near the middle of the data, so it isn't surprising that the mode can go higher even when the median gets lower, and vice versa.

The one thing that is worth noting is that on Friday the median was quite a bit higher than it was for the rest of the week. This doesn't mean much by itself but it does suggest that perhaps people generally took a longer break on Friday.

Keep a note of this. We'll be coming back to it later.

2. Here are the results for the two classes of students taking the multiple choice test.

Table 8.9: Measures of location for multiple choice test data

Set	Median	Mode	Modal class
A	18	2	10–19*
B	27	3	40–49*

*Remember that the modal class depends on how the data was grouped. We had other answers for these as well.

We've already seen that the mode doesn't tell us much for this type of data. In fact, most of the values we could use for the mode aren't even in the modal classes.

We can see that the median for set A is a lot lower than for set B. Taken together with the modal class, it seems as if set B did better than set A.

Question 8.4: Finding the mean of a set of data

Table 8.10: Measures of location for lunch break time data

Day	Mean	Median	Mode
Monday	63.45	62	62
Tuesday	62.9	61.5	65
Wednesday	64.7	62	62
Thursday	64.05	64	67
Friday	72.2	71.5	66
Whole week	65.46	63.5	62

The means follow a very similar pattern to the medians.

Note that once again, it's difficult to interpret the mean. All the break times were given in whole minutes, but the means are all decimals. Nobody was actually away for 64.7 minutes.

You can see one of the useful things about the mean: it makes it very easy to compare data sets that otherwise look very similar.

Question 8.5: Comparing the means of two sets of data

Table 8.11: Measures of location for multiple choice test data

Set	Mean	Median	Mode	Modal class
A	20.88	18	2	10–19*
B	27.56	27	3	40–49*

*Remember that the modal class depends on how the data was grouped. We had other answers for these as well.

Again, the mean values don't entirely make sense as actual marks, but they seem to confirm what the median suggested, that set B did better than set A.

Question 8.6: Extreme values

Remember that the mode and median look for a single value within the data set. Even when the median lies between two values, it is only affected by those two values.

The mean is found by using every single value in the data set – it is much more representative of the data, but is more likely to be affected by a single extreme value.

Question 8.7: Which measure of location should you use?

Most (over 50%) hadn't read the magazine at all. But we don't know whether they are people who would even be interested in the magazine. You might look more closely at the people who did read it. Of those, most read most of the issues, so the magazine seems to have a high level of loyalty from its readers.

Question 8.8: How useful is the mean?

Obviously, you can only have a whole number of children. But look back at the notes on comparing two means. The numbers don't tell us anything about the actual number of children anyone has, but they do allow us to make a comparison between 'then' and 'now', and the suggestion is that people tend to have fewer children now than they did years ago.

Question 8.9: Finding measures of location for your own data

There's no right or wrong answer for this question, since this is your own work on your own data. However, make sure you keep your work safe, because we'll be returning to it in chapter 9.

8.10 Feedback on activities

Activity 8.1: Finding the mode of a set of data

1. The easiest way to answer this is to construct a frequency table, something you did in chapter 7. If you can't remember how to do this, check back.

The frequency table should look like this:

Table 8.12: Frequency distribution of ages of students on a foundation course

Age of student	No. of students (frequency)	%
18	9	30
19	6	20
20	6	20
21	4	13
22	2	7
23	1	3
24	0	0
25	1	3
26+	1	3
Total	30	100

Note: Percentages do not sum to total due to rounding.

Clearly, the age that occurs most often is 18. Nine students are 18 years old. So for this set of students the modal age is 18.

You probably noticed that we didn't give the frequencies for all the ages between 26 and 35. we could have done, just as we could have given them for ages 37 to 65. They'd all have had a frequency of zero.

When we have just a few values a long way away from the main data, it's quite common to do this to make life a little easier. This is called an open class. We have to be careful what we do with it – we mustn't forget what it really represents.

N.B. It is a very common mistake when asked for the mode, to look for the highest frequency and give that as the answer. In this example, the highest frequency is 9. But that isn't the mode. The mode is the thing that happens nine times – in this case, that means the age of 18.

2. What about the second question? Can we use the mode to say how old most of the students are?

Usually, when we use the term 'most', we mean more than half. So are more than half of the students aged 18? No. In fact, it would be accurate to say that most of the students are aged 19 or above, even though 18 is the single age that occurs most often.

Activity 8.2: Finding the mode 2

1. In this case most of the numbers appear only once or twice. The only one that appears three times is 1, so the mode is 1 (not 3!).

2. In this case there are so many different data values, with each one appearing quite rarely, that the mode doesn't really give a good idea of the data. For this sort of data it may be better to group the data, as we'll see.

Activity 8.3: Finding the modal class for a set of grouped data

You can group the data either in groups of ten or five.

If you chose to put the scores in groups of ten, you should have got either of the following:

Table 8.13: Frequency distribution of multiple choice marks (set A)

Mark	No. of students (frequency)	%
0-9	7	22
10–19	11	34
20–29	5	16
30–39	4	13
40–49	5	16
50	0	0
Total	32	100

Note: Percentages do not sum to total due to rounding.

Figure 8.3: Stem and leaf plot of multiple choice marks (set A)

											Count
0	0	2	2	2	2	6	7				(7)
1	0	1	1	3	4	7	7	7	8	8	8	(11)
2	0	1	4	6	7						(5)
3	5	6	7	7							(4)
4	1	3	4	5	7						(5)
5											(0)

With this data set, we had to include zero in the first group, which means that a score of 50 would have to go in a group of its own. But since nobody got all 50 questions right, that isn't something we have to worry about here.

If you group the data this way, the modal mark for the test is between 10 and 19.

Putting the data in groups of 5 would give this result:

Table 8.14: Frequency distribution of multiple choice marks (set A)

Mark	No. of students (frequency)	%
0–4	5	16
5–9	2	6
10–14	5	16
15–19	6	19
20–24	3	9
25–29	2	6
30–34	0	0
35–39	4	13
40–44	3	9
45–49	2	6
50	0	0
Total	32	100

Figure 8.4: Stem and leaf plot of multiple choice marks (set A)

							Count
0	0	2	2	2	2		(5)
0*	6	7					(2)
1	0	1	1	3	4		(5)
1*	7	7	7	8	8	8	(6)
2	0	1	4				(3)
2*	6	7					(2)
3							(0)
3*	5	6	7	7			(4)
4	1	3	4				(3)
4*	5	7					(2)
5							(0)

Now we get a different result. The modal score is between 15 and 19. Note that even though we only have scores of 17 and 18, the group they are in is for 15 to 19, so that's what we give as the modal class.

In this example, the modal class in the second method was a sub-group of the modal class in the first, but this isn't always the case.

Activity 8.4: Ranking a set of data

Here is a stem and leaf plot of the data. It doesn't matter if you used different groupings, as long as the data is in the right order and you have included the count.

Figure 8.5: Stem and leaf plot of ages of students on a foundation course

																Count
1																(0)
1*	8	8	8	8	8	8	8	8	8	9	9	9	9	9	9	(15)
2	0	0	0	0	0	0	1	1	1	1	2	2	4			(13)
2*	1															(1)
3																(0)
3*	6															(1)

The stem and leaf plot puts the data in order, so it's easy to write the data as a ranked list from this.

Keep a note of this. We're going to use this to find the median value later.

Activity 8.5: Finding the median for a set of data

You could have used either of the plots. We'll use the one with groups of 5.

Figure 8.6: Stem and leaf plot of multiple choice marks (set A)

							Count
0	0	2	2	2	2		(5)
0*	6	7					(2)
1	0	1	1	3	4		(5)
1*	7	7	7	8	8	8	(6)
2	0	1	4				(3)
2*	6	7					(2)
3							(0)
3*	5	6	7	7			(4)
4	1	3	4				(3)
4*	5	7					(2)
5							(0)

We have 32 students, so $n = 32$.

The position of the median is found by:

$$\frac{n+1}{2} = \frac{32+1}{2} = \frac{33}{2} = 16.5$$

So we need to find the number between the 16th and 17th values.

Using the count, the first three rows add up to 12 items. The next row has a count of 6, so the 16th item is the fourth number in that row, which equals 18. The next item, the 17th, also equals 18, so we don't have to do any more work. The median is 18.

Activity 8.6: Finding the median 2

If we draw a stem and leaf diagram for this data we get:

Figure 8.7: Stem and leaf plot for email data

									Count
0	0	1	1	1	2	3	3	4	(8)
0*	7	8	9						(3)
1	0	1	2	3	4				(5)
1*	5	6	6	9					(4)
2	1	2	4						(3)
2*	9								(1)

We have 24 people, so the position of the median is:

$$\frac{24+1}{2} = 12.5$$

The median is therefore between the 12th and 13th numbers. Using the counts, the first two rows contain 8+3 = 11 numbers. So the 12th and 13th numbers are the first two on the third row. These are both 10 and 11, so the median is halfway between 10 and 11, so the median is 10.5.

Activity 8.7: Finding the mean of a set of data

1. You could have worked out the mean by adding all the numbers and dividing by 30:

$$\frac{18+22+18+...+19}{30}$$

But it would make more sense to use the information from the frequency distribution:

$$\frac{(18\times9)+(19\times6)+...+(1\times36)}{30}$$

It may not make much difference for a small set of data, but for a large set, it will be quicker this way. Make sure you know how to do this on your calculator – they all work slightly differently from each other.

Either way, you end up with:

$$\frac{608}{30} = 20.27$$

2. The three measures of location are given in the table below.

Table 8.15: Measures of location for student age

Measure	Mean	Median	Mode
Student age	20.27	19.5	18

3. It is easy to explain what the mode means. It's the most common age among the students in the class.

   The median also makes sense. It is the value that's right in the middle of the data. There is not a student with that age, but it is a value that is right between the ages of two students.

   The mean is more difficult to interpret. There is not a student aged 20.27. It is a number which is derived from the entire data set (which in one sense makes it a good representative of the complete set) but it is a number that does not exist within the data.

4. The three measures of location are given in the next table.

Table 8.16: Measures of location for exam mark data

Measure	Mean	Median	Mode
Exam marks (Ch 7, Act. 7.10)	53.58	53.5	63

5. You should have got something like figure 8.8 below:

Figure 8.8: Bar chart of ages of students on a foundation course

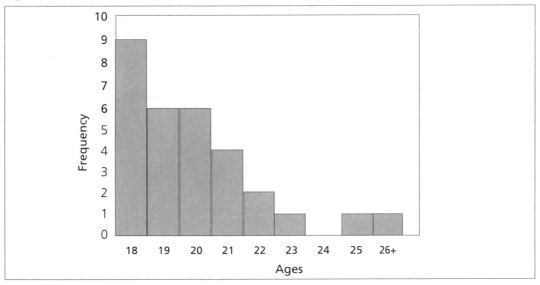

The histogram of the exam marks in chapter 7 was symmetric, but this isn't. It's stretched out. Try marking the mean, median and mode on both sets of data.

Figure 7.16: Histogram of examination marks

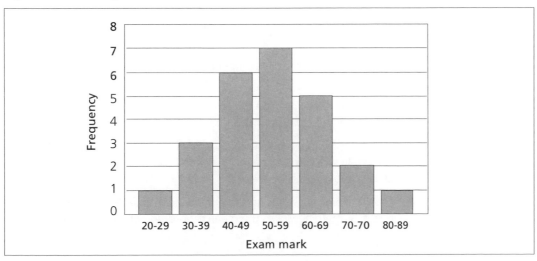

The mean and median are very close in the exam marks data. The mode is different, but the modal class includes the other two, so we have three measures of location very close to each other.

We will be looking at the shape of data in chapter 9, but in the meantime think about what the three measures of location might tell us about the shape.

Activity 8.8: Finding the mean 2

1. To find $\sum x$ we just have to add up all 24 data items. We find that $\sum x = 261$. The mean is then
$$\frac{\sum x}{n} = \frac{261}{24} = 10.875$$
Since we can only have a whole number of emails, once again the mean does not represent an actual number that could be in the data. However, it still provides a useful measure of where the centre of the data is.

2. The three measures of location are shown in table 8.17:

Table 8.17: Measure of location for email data

Measure	Mean	Median	Mode
No. of emails	10.875	10.5	1

The three measures are rather different. The mean is greater than the median, and the median is greater than the mode. This is because the data are not very symmetrical. When we group the data the first few groups are large and the later groups are small, as happened with the age data in activity 8.7. The technical name for this is *skewed data*, and we'll see more examples in chapter 9.

Activity 8.9: Using the mean to differentiate between two data sets

Here are the values you should have got:

Table 8.18: Differentiating between two data sets

Measure	Mean	Median	Mode
Start of day	1.3	1	1
End of day	1.8	1	1

In both cases, the most common number of faulty components in a box was 1 and so was the middle value.

The values for the mean don't really make sense on their own. We can't possibly have 1.3 faulty components. Either a component is faulty or it isn't.

But look at the data again. In the morning, we had nine boxes with no faulty components, but by the end of the day we have only six. Also, in the morning, no box had more than four faulty components, but later on, two boxes have five.

From the data, it looks like the machine is producing more faulty components at the end of the day than it did at the beginning, something which the mean picked up, while the median and mode didn't.

CHAPTER 9

Summary statistics 2 – measures of spread

OVERVIEW

In the last chapter we considered three different measures of location which we could use as summary statistics.

The three measures – the mean, the median and the mode – all had specific purposes and were useful in different situations, but none of them was universally useful on its own. Sometimes it helped to consider two, or even three of them together, but even this didn't always seem to be a really effective way of describing a complete data set.

In this chapter we will continue to use these summaries, but we're going to try putting them in the context of the whole data set. Obviously, we can't just describe the whole of a data set – the whole point of summary statistics is to save us the effort of presenting the whole lot. But what we can do is describe how the data is spread out away from the centre of location. We'll do this by measuring the distance the data covers, and we'll also consider the pattern or shape of the data.

Learning outcomes	On completion of this chapter, you should be able to:

- Choose an appropriate measure of spread for a particular measure of location

- Calculate the value of a measure of spread

- Interpret the measure of spread in the context of its associated measure of location

- Recognise the limitations and strengths of each measure of spread, singly and in combination with its measure of location

- Choose an appropriate set of summary statistics for a given situation

- Use a combination of measures to draw conclusions about the shape of a set of data.

9.1 **Recapping measures of location**

In the last chapter, we tried to find a way of summing up a complete data set by using just one number. Well, actually, we ended up using three. These were the measures of location – or measures of central tendency – called the mode, the median and the mean.

The mode found the most common value in a set of data. This was sometimes very useful, but often was not. The mode wasn't necessarily typical of the rest of the values, and sometimes didn't tell us anything helpful, if it existed at all. Also, sometimes we had to settle for a modal group or class, but that may or may not have included the mode, and would be different depending on how the data was grouped.

The median should have been a more representative summary measure, since it was the value that was right in the middle of the data once it had been put into order. Because it's found purely by position, it isn't affected by what happens to the data around it, which makes it a stable value, but also makes it insensitive to what's happening to the data as a whole.

In some ways, the mean seems to be similar to the median, in that it is a kind of middle value. It's found by a sort of 'sharing out' process, so that it does, in some way, reflect the entire data set. But again, that can sometimes be a bad thing, because most of the data can be very close together, and one very different, or *extreme* value can have a drastic effect on the mean. Of course, the mean is often a number that makes no sense in the context of the data. It is, though, a very good way of comparing two or more apparently similar data sets.

You might also have noticed a suggestion that plotting the mode (or modal class), median and mean on a graph can give some idea of how the data is spread out, and what shape it makes. We will come back to looking at shape later, but for now we're going to start looking at ways of describing how data is spread.

9.2 **Measures of spread**

The mode is a tricky thing to deal with. It tells us one thing and one thing only: if we had to guess at the next result from this data, what would our best guess be? To put this in context, we would, perhaps, have to see the whole of the data, or perhaps just a frequency distribution would do. But remember that we're trying to sum up a set of data so that we don't have to give all this information, or draw lots of diagrams and charts.

The median and the mean are a little more versatile. They can often be accompanied by other numbers that tell us how close they are to the data set as a whole. These numbers are sometimes called *measures of dispersion* (mostly by the same people who call measures of location *measures of central tendency*), but we're going to call them measures of spread. They mean the same thing, but 'spread' is easier to remember.

The mean and median each have their own measures of spread. Well, actually they have two each, and we're going to look at them, and why they're useful (as well as when they're not), in the rest of this chapter.

Before continuing, make sure you've worked through the whole of chapter 8. We'll be using the same data sets, including the one you collected about your friends or fellow students. This chapter also builds up as you go along, with the later parts using the earlier parts, so it's important to work through it carefully in order.

9.3 Data sets for this chapter

Here are the main sets of data we used in the last chapter:

Data set 8.1: Ages of a hypothetical set of students

18	22	18	19	25	20	19	18	20	22
21	19	18	20	20	36	19	21	20	18
19	18	18	21	23	18	21	18	20	19

Data set 8.2: Number of emails sent in a hypothetical company

1	12	22	8	0	16	24	4
14	3	1	11	13	29	3	9
10	21	19	2	15	1	16	7

Data set 8.3A: Multiple choice test marks (set A) out of 50

2	37	41	2	18	2	14	7
2	36	11	43	26	37	17	27
24	17	44	20	18	6	35	13
17	18	47	21	0	10	45	11

Data set 8.3B: Hypothetical multiple choice test marks (set B) out of 50

3	28	3	3	50	1	42	1
14	5	48	12	33	21	43	50
24	50	3	39	19	44	43	21
47	21	49	49	47	28	26	15

Data set 8.4: Time taken for lunch breaks (in minutes)

Monday:

47	54	58	58	58	59	60	61	62	62
62	62	63	64	64	65	65	76	82	87

Tuesday:

50	52	54	56	58	59	59	60	61	61
62	63	65	65	65	66	68	74	80	80

Wednesday:

51	53	58	58	59	59	60	61	62	62
62	63	64	65	70	72	73	80	81	81

Thursday:

48	52	55	58	60	60	62	62	63	64
64	66	67	67	67	69	70	72	77	78

Friday:

15	61	62	63	65	66	66	66	69	71
72	72	74	79	84	85	86	86	95	107

9.4 Measures of spread for the median

The range

The median is found by putting the data in order and finding the middle value, so it makes sense that the measures of spread should also be found from ranked data.

The first measure of spread is called the *range* and to find this we need to find the minimum and maximum values in the data. The range is simply the difference: maximum minus minimum.

Finding the range for a set of data

1. Using data set 8.1, find the minimum and maximum for the student ages data. Then use these to find its range.

2. Do the same for data set 8.2.

Problems with the range

The range is a good and easy way of putting the median into context, especially as the minimum and maximum are usually given as well.

But sometimes the range can be affected by just one value that's very different from the rest of the data. We've already come across this problem when we looked at the mean. One extreme value, which we often call an *outlier*, can have a really huge effect, and because we're trying to avoid having to look at the whole set of data this kind of thing could be easily missed.

Have a look at the students' ages data again.

The mode was 18, the median 19.5 and the mean 20.27. They are not all the same, but they are pretty close together.

The range is 18, which suggests that there's a wide variety of ages in the class, but that is not true. Most of the students are aged between 18 and 21, with just four aged 22 to 25. Therefore nearly all the students are in a range of 7. But one older student, aged 36, drags the range from 7 to 18, suggesting a much greater variability in age than is true for most of the students.

Of course, I'm not suggesting that we shouldn't know the full range of the data. Sometimes that is exactly what we want to know. However, sometimes what we really want is to know about how most of the data clusters around the median, so we need another way of describing the spread around the median which will not be affected by one or two extreme values.

Quartiles

In chapter 8, we found the median by ranking the data and splitting it in two to find the middle value. We used the formula

$$\frac{n+1}{2}$$

to find the position of the median.

Now we're going to split the data into four equal parts, and the points in the data that divide it are called *quartiles*.

Here is some ranked data (don't worry about what it is for: we're just using it to find the quartiles).

| 1 | 3 | 4 | 7 | 8 | 10 | 11 | 13 | 14 | 16 | 17 |

Here $n = 11$, so the position of the median is $\frac{n+1}{2} = \frac{11+1}{2} = \frac{12}{2} = 6$.

The sixth data item is in the middle. So the median is the number in the sixth position, which is 10.

The 1st quartile

The 1st quartile (also called the *lower quartile*) splits the bottom half of the data into two, and to find its position we use the formula

$$\frac{n+1}{4}$$

So the position of the 1st quartile is:

$$\frac{n+1}{4} = \frac{11+1}{4} = \frac{12}{4} = 3$$

The number in the 3rd position is 4, so 4 is the 1st quartile.

The 3rd quartile

The 3rd quartile (also called the *upper quartile*) splits the top half of the data into two, and to find its position we use the formula:

$$3 \times \frac{n+1}{4}$$

So the position of the 3rd quartile is:

$$3 \times \frac{n+1}{4} = 3 \times \frac{11+1}{4} = 3 \times \frac{12}{4} = 9$$

The number in the 9th position is 14, so 14 is the 3rd quartile.

You might be wondering what happened to the 2nd quartile. Well, that's simply the median.

The lower quartile is sometimes called Q_1 and the upper quartile is sometimes called Q_3.

Activity 9.2

Finding quartiles 1

Seven friends discuss their ages, which are as follows:

| 18 | 20 | 20 | 21 | 22 | 24 | 24 |

Find the median and the lower and upper quartiles for this data.

Finding the quartile values

In these examples it was easy to find the values of the quartiles, because the formulas produced whole numbers each time. But we already know that sometimes the median falls in between two values, and very often so do the quartiles.

When the median falls between two values, simply average them by adding them together and dividing by two. The quartiles are a little more complicated, and in fact there are several different methods of finding the quartile values.

To keep things simple, if the quartile position is not a whole number, we're simply going to round it to the nearest whole number.

Activity 9.3

Finding quartiles 2

Find the quartiles for the email data in data set 8.2.

Five-figure summaries

The median is often used on its own, but it can also be used along with other measures of location or its associated measures of spread – the range and the positions of the quartiles.

The measures of spread, however, are never used on their own. They don't mean anything without a comparison to the median.

For example, suppose you found out the cost of a cup of coffee in all the restaurants and coffee shops in a busy street. The range for the price of coffee is £2. That might seem quite a big difference, especially if the median price is £1.50. A cup of coffee might then cost between around £0.50 and £2.50, with no guarantee that the most expensive cup tastes any better than the cheapest. Although, you would hope it would. But, what if the median price was £4? The price is then ranging from around £3 to £5. Somehow, the difference doesn't seem to be quite so extreme, even though the range is the same. It also seems to be a very expensive place to buy coffee.

Because the median is quite limited in what it tells us, and everything else doesn't really make sense without the median, it's quite common to give summary statistics as a *five-figure summary*.

We originally wanted one number, then thought we might get away with three, and now we are up to five. (And, just in case you were wondering, no, we're not going to stop there.) But don't panic! There's nothing in the five-figure summary you haven't already seen.

The five-figure summary is just a list, in order, of the minimum, lower quartile, median, upper quartile and maximum. It's often presented like this:

Table 9.1: Five-figure summary for student ages data *(data set 8.1)*

Minimum	18
Lower quartile (Q_1)	18
Median	19.5
Upper quartile (Q_3)	21
Maximum	36

Activity 9.4

Five-figure summaries

Write down the five-figure summary for the email data in data set 8.2.

The point is that the five-figure summary preserves the kind of details about the data that you may want to know about. For instance, even though we only know a few numbers, this is enough to give us an idea of the shape of the data. And, if you can't see that from the numbers in the table, there's a kind of graph that can help you – the box and whisker plot.

The box and whisker plot

The idea of a box and whisker plot is simple. We draw a graph, a bit like a bar chart, with only one axis. This axis represents the values in the data set. Then we draw a box with one end at each quartile, and a line through the middle where the median is. Then we draw a little line or whisker at the minimum and maximum values, and join them up to the box.

Figure 9.1: Construction of a box and whisker plot
Some people don't think the 'whiskers' look much like whiskers, so they just call these *boxplots*.

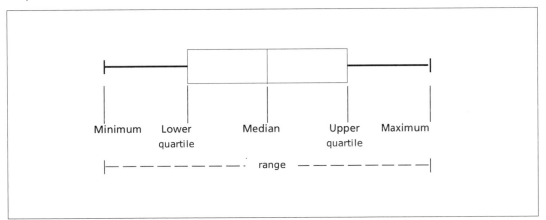

By the way, this boxplot, and the one in figure 9.2, are drawn horizontally. But there's no rule about this. You may prefer to draw them vertically. It's really up to you, and will probably depend on the range you have to cover and the size of your paper (or screen, if you do this on a computer).

Activity 9.5

Using a five-figure summary to draw a boxplot and wisker plot

1. Using a five-figure summary for the student ages data, draw a boxplot.

2. Using a five-figure summary for the email data, draw a boxplot.

Comparing data sets using a box and whisker plot

So far, we've just looked at plots of a single set of data. But we often use them to compare two sets of data directly. There's no problem with this; you simply draw two (or more) plots, one on top of the other, or side by side if you're drawing vertically. Just make sure the scale you use is sensible for both sets.

The reason this is so useful is that you can compare sets of data of different sizes. Remember from chapter 7 that this can be a problem.

Suppose we count how many people are off sick one day in two different departments of a company. You're told that there are 5 people off in department A, and 10 people off in department B. Clearly, there are twice as many people off sick in department B. But does that mean that there's a problem with the staff in department B?

What if you now discover that there are 30 people in department A, and 100 people in department B? Now it seems perfectly reasonable that there are more people away in department B – it's a much bigger department.

The solution to this problem is to standardise your data, which is usually done by turning everything into percentages.

A box and whisker plot doesn't quite turn everything into percentages, but what it does is quite similar. It splits the data into quarters, so while each section of the plot may represent a different number of values or data items, it still represents 25% of the data set.

Figure 9.2: Breakdown of data in a box and whisker plot

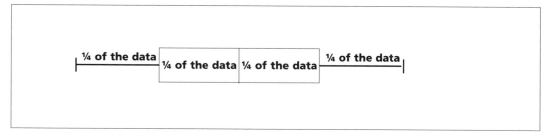

Without knowing the size of each data set, we can now compare them fairly.

Box plots

Go back to the data for students taking the multiple choice test (data sets 8.3A and 8.3B).

1. Find the positions and values of the 1st and 3rd quartiles for the two sets.

2. Combine the summary statistics you have for the multiple choice test data into five-figure summaries for each set of students.

3. Draw a box and whisker plot of both sets on the same axis so that you can compare the groups' performance.

The interquartile range

Once we've found the quartiles, we're ready to find an alternative to the range – the *interquartile range* (or IQR for short).

The interquartile range is found by subtracting the lower quartile from the upper quartile. It ignores the most extreme half of the data, and gives the range of the middle half.

Activity 9.7

Finding the quartiles and IQR for a set of data

Using data set 8.1, find the upper and lower quartiles for the student ages data. Then use these to find the interquartile range.

Activity 9.8

IQR 2

Go back to the data for students taking the multiple choice test (data sets 8.3A and 8.3B).

1. For each set of data find the range and the interquartile range.

2. Compare what the two measures of spread tell you about the two sets of data.

9.5 Measures of spread for the mean

The mean is the measure of location with which you're probably most familiar. Although we often talk about the *average*, what we're really talking about is the mean, rather than any other measure of location.

We're so used to talking about and using the mean that it's easy to forget its limitations. We know that the mean can often be a number that doesn't really make sense – like a family having 2.4 children. We also know that if most of the data in a set is very close but we have just one extreme value. Then, whether it's very much higher or lower than the rest, it can have a large effect on the value of the mean.

So it's very important to know how widely spread the data is. We want to know the average *deviation* from the mean. That's simply how far away from the mean we would expect a randomly chosen data value to be.

We actually have two commonly used measures of spread, called the *variance* and the *standard deviation*. They're very closely linked to each other – so close, in fact, that if you know one, then you can work out the other. However, to understand what they are, and how we can find them, we're going to have to go through quite a detailed explanation.

You don't necessarily need to be able to explain the next few sections yourself. They are just here to help you understand what's going on. But it may help you if you work through the examples yourself as we go along.

The average deviation

Since we want to know how far away, on average, each data item is likely to be from the mean, it makes sense to find all the differences and find their mean.

That sounds really confusing, so let's look at an example.

Five people in a room have different amounts of money in their pockets: here is the data. It has been ranked to make the example a little easier to follow.

£1 £2 £3 £4 £5

The mean value is found by adding up the money and dividing by the number of students, or:

$$\bar{X} = \frac{\sum x}{n}$$

Here, that gives a mean of £3. If we shared out the money equally, everybody would get £3.

To find out how widely spread the data is, though, we need to find out how far, on average, each person's money is from the mean. So we can subtract the mean from each amount of money. That's what we call the *deviation*. Then we just have to find the average of the deviations, and we already know how to do that – add them up and divide by how many there are.

So here goes. We'll call the original amounts of money x, and the deviations are simply x minus the mean.

Table 9.2a: Finding the average deviation for a set of data

	x	$x - \bar{x}$
	1	-2
	2	-1
	3	0
	4	1
	5	2
$\sum$	15	0

If the mean is $\frac{\sum \text{data}}{n}$, then the average deviation should be $\frac{\sum(x-x)}{n}$.

But look at $\sum(x-\bar{x})$ in the table.

It's equal to zero. And, if you think about it, that's always going to happen. The whole point of the mean is that it's in the middle. Some data items will be higher, some lower, and the differences will cancel out.

The problem is, we want to know how far away each data item is from the mean, but we don't want to know if it's higher or lower. In other words, we want the distance, but don't care about the direction.

The mean deviation

This time, we're going to work out the deviations, but we're not going to bother whether they're positive or negative. We're going to take the *absolute value*, or *modulus*, of the deviations. The symbol for this is two lines, one on either side.

Table 9.2b: Finding the mean deviation for a set of data

	x	$x - \bar{x}$	$|x - \bar{x}|$
	1	-2	2
	2	-1	1
	3	0	0
	4	1	1
	5	2	2
Σ	15	0	6

The mean deviation is simply the average of the absolute deviations, or $\dfrac{\sum |x - \bar{x}|}{n}$, which is $\dfrac{6}{5} = 1.2$

There really is nothing wrong with that. The mean deviation is a perfectly legitimate measure of spread for the mean. There really is only one problem with it – nobody ever uses it!

The reason is that the modulus, or absolute value, is very easy to do, but very difficult to work with mathematically. Even though it serves its purpose here, it's just not considered a very good thing to do unless there is no better alternative. Fortunately for us, there is.

The variance

There is another way to make sure that every number we use is positive. There is one thing we can do to any number, positive or negative, so that the result is always positive – multiply the number by itself. If you square a positive number, you get a positive number, and if you square a negative number you still get a positive number.

So let's try again.

Table 9.2c: Finding the variance for a set of data

	x	$x - \bar{x}$	$(x - \bar{x})^2$
	1	-2	4
	2	-1	1
	3	0	0
	4	1	1
	5	2	4
Σ	15	0	10

We are still going to add up our (squared) deviations, and divide by how many there are, so now we have $\dfrac{\sum (x - \bar{x})^2}{n}$ or $\dfrac{10}{5} = 2$

This is the variance, which is also a perfectly legitimate measure of spread. This time, it is actually used – it's considered a very important summary statistic, and lots of statistical techniques use it.

So you might think that's it. However, there is still a problem with the variance. If you think about what we've done, you will realise that the value of the variance is quite big. It's certainly bigger than the mean deviation (in this case), and the reason is obvious. We squared all the deviations. That means that any deviations below 1 will get smaller, and any deviations above 1 will get bigger. So how can we be sure that the variance makes sense?

There is also another problem which might seem silly at first, but is actually very important.

You probably know how to work out areas of things. So if you have a field that's 300 metres long and 200 metres wide, you find the area by multiplying them together.

300m x 200m = 60,000m². But look at what we did. We didn't just multiply the numbers together; we also multiplied the units, so that instead of metres, the area is given in square metres (m²).

Now look at the numbers we used in our example. They were not just numbers; they were amounts of money, and the deviations were also amounts of money. So the squared deviations weren't pounds, but square pounds (£²).

This is a bit of a problem, because it's not really clear what a square pound is, or whether it even means anything.

This may seem like a lot of fuss over nothing, but the mean was a quantity of money, measured in pounds, and the measure of spread really only makes sense if it is in the same units as the mean.

So we are going to have to find a way of undoing all the squaring we did.

The standard deviation

The simplest thing we can do now is to take the value we got for the variance, complete with its squared units, and take the square root of the whole thing. That gives us the standard deviation, which we usually abbreviate to the symbol s.

So $s = \sqrt{\dfrac{\sum (x - \bar{x})^2}{n}}$ and thus in the money example, $s = \sqrt{2} \approx 1.41$

Notice that s is quite similar to the mean deviation.

You should also have noticed that, since the standard deviation is the square root of the variance, the variance must be the square of the standard deviation. So the symbol used for the variance is s^2.

That was a lot of hard work! But now, all you really need to know is how to work out a standard deviation for yourself.

Activity 9.9

Finding the variance and standard deviation for a set of data

1. Using data set 8.1, find variance and standard deviation for the student ages data.

2. Using data set 8.2, find the variance and standard deviation for the email data.

An important note about the standard deviation

This book is a foundation course, and because a lot of people find mathematics and statistics a bit scary, we are trying to make things a bit easier. But sometimes things crop up that are important to know about, even if you don't necessarily understand them fully. And this is one of them.

In this chapter we've seen how to calculate the variance and the standard deviation. You don't necessarily need to be able to explain the formulas, but you do need to be able to use them to calculate the variance and the standard deviation.

However, in the real world it's a bit more complicated than that. We've seen some basic ways to explore data, but the variance and standard deviation can be used in several more advanced ways to explore sets of data in more detail. We've used the formula

$$s^2 = \frac{\sum (x - \bar{x})^2}{n}$$

for the variance, since we just wanted the mean squared deviation. However, in some situations it turns out that it's better to use the formula

$$s^2 = \frac{\sum (x - \bar{x})^2}{n-1}$$

for the variance, so instead of dividing by n we divide by $n - 1$. This will change the standard deviation as well, of course.

To be honest, if n is a big number then it doesn't really make much difference whether we divide by n or $n - 1$. The reasons why we sometimes use one formula and sometimes use the other are quite subtle, and a full explanation would be way beyond the scope of this book. For the kinds of problems we've looked at in this module you can stick to the n version all the time.

The reason we are mentioning the other version is that you'll come across it in lots of textbooks on statistics, and if you discuss statistics with other people. Also, some calculators and computer packages can calculate variances and standard deviations, and they usually have two different buttons or functions so that you can calculate both versions.

Try calculating the standard deviation for data sets 8.1 and 8.2 using the $n - 1$ formula, and compare the values with your results from activity 9.9.

The coefficient of variation

When we worked out the interquartile range, it didn't make much sense on its own. We needed to put it together with the median for it to mean anything.

We have the same problem with the standard deviation.

Suppose we knew the earnings of ten people, and we found out the standard deviation for their earnings was $100,000. That is an awful lot of money, but what does it tell us about the ten people? Well, not much really. We might have a lot of people on low incomes, and one millionaire. Or we might have ten highly paid sports stars. We simply don't know what this standard deviation means in real terms without knowing something about the mean.

Or, suppose we have a machine producing rubber balls. We measure the diameter of the balls to find their standard deviation, and it is 0.2mm, which is a very small difference. If we are producing footballs, with a mean diameter of around 30cm, this is a very small difference, which nobody would ever notice. But, if we are producing rubber bearings for use in specialist medical equipment, and they have a mean diameter of 1mm, then 0.2mm is quite a large difference.

So we need some way of comparing the standard deviation with the mean to get some idea of how important it is. One final number (sorry, but we did say we weren't finished), which we call the *coefficient of variation,* does exactly that.

The good news is that this is very simple to work out, because it is just the standard deviation divided by the mean, then multiplied by 100.

Activity 9.10

Finding the coefficient of variation for a set of data

1. Using the mean and standard deviation already calculated, find the coefficient of variation for the student ages data.

2. Similarly, find the coefficient of variation for the email data.

9.6 Choosing measures of location and spread

This seems like a good place to stop and go over what we've done in the last two chapters. The next couple of sections should just be a reminder, so if there's anything that seems new or unfamiliar to you, go back over the earlier sections.

Measures of location

To begin with, we wanted to find some way of summing up a complete data set without having to give all of the data, or drawing complicated graphs and diagrams.

We actually came up with three different ways to do this, called measures of location. All three of them, the mean, the median and the mode, told us slightly different things about the data, and all had their uses, but none of them was good in every situation – it all depended on the data itself, and what we wanted to know.

Even though none of the measures of location was useful in every situation, looking at all three together could sometimes be useful if we wanted to describe a set of data in a different way – by its shape.

If a graph of a set of data seemed to be fairly symmetrical, we could expect that the mean, median and mode would all have very similar values. Of course, since we're trying to avoid making lots of graphs, it's useful to know that we can do this the other way around. If the three measures of location are very similar in value, then the data is likely to have a symmetrical shape.

So what about if the three measures of location are quite different? Well, then we expect the data to be bunched up at one end, and stretched out at the other end. We call this shape *skewed.* The data can be very bunched, or just a little bit bunched. It can have a long tail or a short tail. And, it can be skewed in either direction.

We can usually tell which direction data is skewed in by looking at the measures of location.

Figure 9.3a: Measures of location for a set of data skewed to the left (negatively skewed)

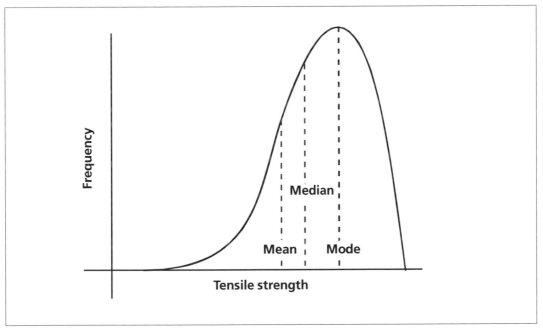

If the measures of location are in alphabetical order – mean, median, mode – then the data has a negative or left skew. But if they're in reverse alphabetical order – mode, median, mean – then the data is positively skewed.

Figure 9.3b: Measures of location for a set of data skewed to the right (positively skewed)

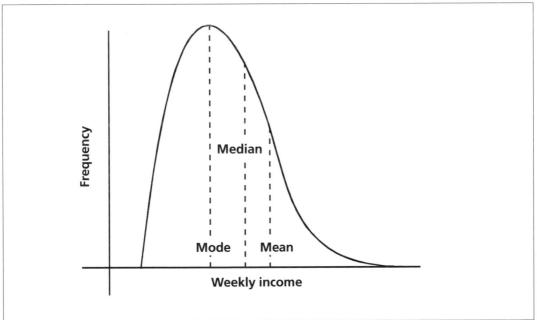

Even though the individual measures of location – and the three together – can tell us useful things, there are still things about a set of data that they don't tell us. So we also need to know how widely the data is spread.

Measures of spread

Measures of spread help to put the measures of location into some kind of context and help to understand what the measures of location mean. The measure of spread you use depends on which measure of location you're using. They are not interchangeable.

Deciding what to use

So which method, or combination of methods, should we use on a set of data?

There is no hard and fast rule – it all depends on the data, and what you want to know.

But here are some guidelines. Remember, these are not rules, just things for you to think about when you're trying to decide how to treat data:

1. If you're trying to find the shape or pattern of a set of data, comparing the mean, median and mode will help you decide if it's symmetrical or skewed.
2. If you want to estimate the most likely outcome (or next event) from a set of data, use the mode or modal class.
3. If you want to make comparisons between two similar sets of data, the mean is more sensitive to small differences than the other measures of location.
4. The five-figure summary allows you to see the shape of a set of data, and it makes it easy to see the presence of outliers. It's also a good way of comparing sets of data, even if they have different numbers of elements. It also gives you the choice of whether to use the range or the interquartile range. Because of this, this method is often favoured when looking at skewed data distributions.
5. The mean and standard deviation don't allow you to see the shape of the data, but are used to compare different sets of data, and the coefficient of variation allows you to compare the variability between them.

   This method is particularly favoured when working with symmetrical data distributions, or when you want to ignore outliers and concentrate on the central part of the data.

So now it's over to you!

9.7 Summary

In this chapter we continued to look at measures of location and dispersion, and saw how to decide which of the methods to use. We also saw ways to use diagrams to represent these measures clearly. These ideas are important for getting a simple picture of a set of data, so that we can tell someone else the important features of our data, or compare two different sets of data.

9.8 Review questions

 Question 9.1: Finding the minimum, maximum and range for a set of data

Find the data for lunch break times (data set 8.4).

Using ordered data, find the minimum, maximum and range for the sets.

Are the results very different from each other?

Do these results give any indication of what you found in the last chapter?

Can you think of a way of modifying this technique so that it is more representative of most of the data?

 Question 9.2: Finding the upper and lower quartiles and interquartile range for a set of data

Find the interquartile range for each day of the lunch break times data (data set 8.4).

Compare these results with those for the range.

Comment on what you find.

 Question 9.3: Using a five-figure summary to draw a box and whisker plot

Use what you have already done with the lunch break times data to make five-figure summaries of each day.

Use these to draw boxplots so you can compare the five days with each other.

 Question 9.4: Finding the variance and standard deviation for a set of data

1. Find the variance and standard deviation for the two sets of the multiple choice test data (data sets 8.3A and 8.3B).

   Compare the standard deviations with the interquartile ranges for the same data.

2. Find the variance and standard deviation for the lunch break times data.

   Compare the standard deviations with the interquartile ranges for the same data.

3. Try to mark where the mean and standard deviations would be on your stem and leaf and box plots for the two sets of data. Do they make sense?

Question 9.5: Finding the coefficients of variation and standard deviation for a set of data

1. Find the coefficients of variation for the multiple choice marks data.

   What do they tell you about the two classes?

2. Find the coefficients of variation for the lunch break times data.

   Compare the results for the different days. Do they help you describe what happens over the week?

3. Compare the coefficients of variation for the two sets of data in parts1 and 2 above.

   Comment on the level of variation within the two data sets.

 Question 9.6: Recapping the three measures of location

Try to describe in words what the mean, median and mode can be used for. Write down how to find their values, and what their limitations are.

Question 9.7: Recapping the measures of spread

Which measures of spread go with the three measures of location? How do you find them? What are their good points and bad points?

Question 9.8: Creating a report

In the last chapter you collected data from your friends, family or fellow students. You should have a set of data for age, height, number of siblings and distance from workplace or learning centre. You should also have started to prepare a report, including the measures of location for your data.

- Update your report to include five-figure summaries, range and interquartile range. Comment on your results and illustrate them with box and whisker plots.

- Add the standard deviation and coefficient of determination. Compare these with the results you got from the five-figure summaries. Do they appear to confirm what the rest of your data suggests? Are there any differences? If so, can you explain them from the data?

- Compare all of your results. Comment on the shape and level of variability of each of the four sets of data. Use this to decide which method (or methods) is the most suitable for working with each data set. When making your decisions, don't just look at the data. Think about the different ways you might want to use the data. Your decisions may depend on what it is you want to know.

- If you can, discuss your decisions with a tutor and other students. They may have different ideas on what the data could be used for, and that may affect how you work with it.

9.9 Answers to review questions

Question 9.1: Finding the minimum, maximum and range for a set of data

The lunch break times data was given to you already ranked, so it's easy to find the results in the following table:

Table 9.3: Minimum, maximum and range for lunch break time data

Day	Minimum	Maximum	Range
Monday	47	87	40
Tuesday	50	80	30
Wednesday	51	81	30
Thursday	48	78	30
Friday	15	107	92

We don't need to worry about the results for the whole week. Since the highest maximum and lowest minimum both occurred on Friday, they will be the same.

Monday has a slightly higher range than the next three days, although we know that in all other respects the first four days of the week seem to be quite similar.

We also know that it seemed that most people took a lot longer for lunch on Friday. If everybody had, we could expect that the range would be similar to the other days – both the maximum and minimum would have increased. But, the range is actually much bigger, because while most people did take longer for lunch, one person took a very short (15 minutes) break.

Once again, a single value at one end of the data has had a big effect on the summary statistics we're trying to use.

Question 9.2: Finding the upper and lower quartiles and interquartile range for a set of data

Table 9.4: Minimum, maximum and range for lunch break time data

Day	1st quartile	3rd quartile	Interquartile range	Range
Mon	58	65	7	40
Tues	58	66	8	30
Wed	59	72	13	30
Thurs	60	69	9	30
Fri	65	85	20	92

Comparing the IQRs to the ranges, it doesn't seem as if most people are taking very different times for their lunch break. The wide ranges are being caused by just a few people who take very long or short breaks each day.

The biggest difference is for Friday. After removing the effect of the most extreme values, there's only 20 minutes difference between the quartile values. This is still a wider variation than we had on the other days, but nowhere near as wide as the range suggested. But, it still looks like people were taking longer for their breaks on Friday.

Question 9.3: Using a five-figure summary to draw a box and whisker plot

1. Your results should look something like this:

Figure 9.4: Box and whisker plot comparing two sets of students' multiple choice test marks

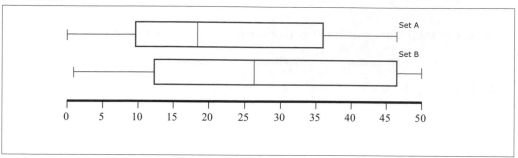

2.

Table 9.5: Box and whisker plot comparing lunch break times

	Mon	Tues	Wed	Thurs	Fri
Minimum	47	50	51	48	15
Q_1	58	58	59	60	65
Median	62	61.5	62	64	71.5
Q_3	65	66	72	69	85
Maximum	87	80	81	78	107

Figure 9.5: Box and whisker plot comparing lunch break times for five days

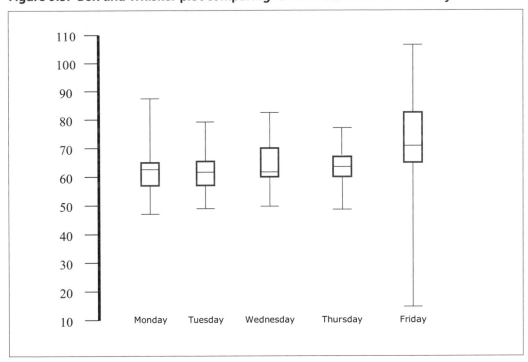

Question 9.4: Finding the variance and standard deviation for a set of data

1. The measures of location and spread for the multiple choice data are given in the table below.

   Using data from chapter 8 we get:

Table 9.6: Measures of spread and location for multiple choice test data

	Mean	Variance	Standard deviation	IQR	Median
Set A	20.88	202.61	14.23	26	18
Set B	27.56	307.93	17.55	35	27

Just like the IQR, the standard deviation is bigger for set B, suggesting the data is a bit more widely spread.

We already knew that set B seemed to score better overall – both the medians and means tell us that – and now we can be pretty sure that the spreads are different as well.

But what about the actual values of the standard deviations and interquartile ranges? They suggest the same things, but the values are quite different.

2. The measures of location and spread for the lunch break times data are given in table 9.7.

Table 9.7: Measures of spread and location for lunch break times data

	Mean	Variance	Standard deviation	IQR	Median
Mon	63.45	77.85	8.82	7	62
Tues	62.90	61.99	7.87	8	61.5
Wed	64.70	72.81	8.53	13	62
Thurs	64.05	53.95	7.34	9	64
Fri	72.20	311.46	17.65	20	71.5

Again, the standard deviations follow the same general pattern as the interquartile ranges, although the actual numbers are different.

3. In this activity, what we are trying to get you to do is to superimpose the mean and standard deviation onto plots usually associated with the median and interquartile range. You may have got reasonable results for some of the plots, but for others, the two sets of summary statistics didn't match up particularly well.

Think about the shape suggested by the box and whisker plots for those that did match up and those that didn't. We will come back to this later. For now, though, it should be clear that you cannot simply mix and match methods of location and spread. The *median* goes with the interquartile range (or range) whereas the *mean* goes with the standard deviation and variance.

Question 9.5: Finding the coefficients of variation and standard deviation for a set of data

1.

Table 9.8: Coefficients of variation for multiple choice marks data

	Mean	Standard deviation	Coefficient of variation
Set A	20.88	14.23	68.19
Set B	27.56	17.55	63.67

We have established that Set B students generally did better than Set A students, but there was also slightly more variability. However, the coefficients of variation are very similar, so any increase in variability is proportional to the increase in overall performance.

2.

Table 9.9: Coefficients of variation lunch break times data

	Mean	Standard deviation	Coefficient of variation
Monday	63.45	8.82	13.91
Tuesday	62.90	7.87	12.52
Wednesday	64.70	8.53	13.19
Thursday	64.05	7.34	11.47
Friday	72.20	17.65	24.45

The coefficients of variation for Monday to Thursday are all very similar, which is not surprising, since they all had similar means and standard deviations.

It also isn't surprising that there is a much higher level of variability on Friday. We already knew the standard deviation was higher than for the other days, but so was the mean. The coefficient of variation tells us that there really was a much greater variability of time taken on Friday.

3. At first glance, it would seem that the multiple choice data is much more variable than the lunch break times, since the standard deviations are higher. What is surprising is how much more variable it is. The coefficients of variation are much higher, because they take into account the means as well as the standard deviations.

Question 9.6: Recapping the three measures of location

All the answers to this review question are in this and the previous chapter. Use these to remind yourself about measures of location.

The main points are:

- **The mode** is the single most commonly occurring value (or range of values), but it may not tell us anything about the data as a whole

- **The median** is the middle value in a set of ranked (or ordered) data. It isn't affected by extreme values – or outliers – which makes it a stable, consistent measure, but also means that in a sense it may not be telling you about the whole of the data

- **The mean** is found by using a formula that includes every single item in the data set, so is representative of the data. But that means it can be affected by outliers that are very different from most of the data. It can also take a value that doesn't really mean anything in context. However, it can show up small differences between two or more data sets that appear to be very similar.

Question 9.7: Recapping the measures of spread

All the answers to this activity are in this chapter. Use these to remind yourself about measures of location.

The main points are:

- The *mode* has no associated measure of spread. If the mode needs to be put into context with the data, it is usually given with the frequency distribution.

- The *median* can either be put into context using the *range* or the *interquartile* range.

  The range is the distance between the maximum and minimum values in the data, which means that it accounts for the whole data set. However, that leaves it susceptible to the effects of extreme values.

  This is overcome with the interquartile range, which ignores everything except the middle half of the data. But that makes it less representative of the whole data set.

  Rather than give both the range and interquartile range, the median is often given as part of a five-figure summary which also includes the maximum, minimum, lower and upper quartiles. These can be used to calculate the range and interquartile range, and can also be used to draw box and whisker plots, which give a good visual description of the shape of the data, and also make it easier to compare data sets.

- The *mean* is usually accompanied by the *standard deviation*, although the *variance* is sometimes used. Since the variance is the square of the standard deviation, they are easily interchangeable.

  Like the mean itself, the standard deviation can be affected by the presence of outliers in the data, and unlike the measurements associated with the median it doesn't give any indication of shape. The variability of data sets with different means and standard deviations can be fairly compared using the coefficient of variation.

Question 9.8: Report

There's no right or wrong answer for this question, since you're working on your own data. However, things to look out for are outliers, data with a high level of variation, and more generally any data that looks interesting or unusual.

Try to compare your results with those of other students. Did you all get roughly similar results? Did any of you find something particularly unusual? Did any of you come up with any interesting conclusions from your analyses?

If possible, try to discuss your results with a tutor. Are they convinced by your conclusions? Can they suggest any ways of analysing the data which you hadn't thought of?

9.10 Feedback on activities

Activity 9.1: Finding the range for a set of data

1. The easiest way to get started with this is to use a stem and leaf plot. You should still have the plot you used in chapter 8.

Figure 8.5: Stem and leaf plot of ages of students on a foundation course

															Count
1															(0)
1*	8	8	8	8	8	8	8	8	9	9	9	9	9	9	(15)
2	0	0	0	0	0	0	1	1	1	1	2	2	4		(13)
2*	1														(1)
3															(0)
3*	6														(1)

From this we know that the minimum equals 18, and the maximum equals 36.

So the range is 36–18 = 18.

2. Using the stem and leaf diagram in activity 8.6 we know that the minimum is 0 and the maximum is 29, so the range is 29 – 0 = 29.

Activity 9.2: Finding quartiles 1

Here there are 7 data items, so $n = 7$.

The position of the median is

$$\frac{n+1}{2} = \frac{7+1}{2} = \frac{8}{2} = 4$$

so the median is the 4th data item, which is 21.

The position of the lower quartile is

$$\frac{n+1}{4} = \frac{7+1}{4} = \frac{8}{4} = 2$$

so the median is the 2nd data item, which is 20.

The position of the upper quartile is

$$3 \times \frac{n+1}{4} = 3 \times \frac{7+1}{4} = 3 \times \frac{8}{4} = 6$$

so the median is the 6th data item, which is 24.

Activity 9.3: Finding quartiles 2

For data set 8.2 there are 24 data items, so $n = 24$.

The position of the lower quartile is

$$\frac{n+1}{4} = \frac{24+1}{4} = \frac{25}{4} = 6.25$$

We can round this to 6, so the lower quartile is the 6th data item, which is 3 (using the stem and leaf diagram in activity 8.6).

The position of the upper quartile is

$$3 \times \frac{n+1}{4} = 3 \times \frac{24+1}{4} = 3 \times \frac{25}{4} = 18.75$$

We can round this to 19, so the upper quartile is the 19th data item, which is 16.

Activity 9.4: Five-figure summaries

We've already found all the bits we need in activity 8.6, activity 9.1 and activity 9.3, so we just have to put them together in the right order:

Table 9.10: Five-figure summary for email data

Minimum	0
Lower quartile (Q_1)	3
Median	10.5
Upper quartile (Q_3)	16
Maximum	29

Activity 9.5: Using a five-figure summary to draw a boxplot and whisker plot

1. You should have found the box and whisker plot for this looks a little different from the one I used as a demonstration.

Figure 9.4: Box and whisker plot for student ages data

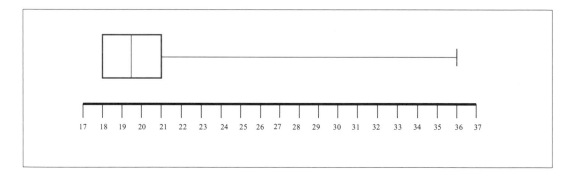

That's because the minimum and lower quartile are the same, so we've only got one whisker.

Bearing that in mind, you can now get a good idea of the shape of the data. It's clearly all bunched up on the left-hand side, with just a little of the data stretched out between the values of 21 and 36.

2. Using the five-figure summary from activity 9.4, we get

Figure 9.5 Box and whisker plot for email data

This plot shows that the data is rather bunched up, but not as much as the student ages data.

Activity 9.6: Box plots

1. The position of the 1st quartile is

$$\frac{32+1}{4} = 8.25$$

which we can round to 8, and the position of the 3rd quartile is

$$3 \times \frac{32+1}{4} = 24.75$$

which we can round to 25.

So, looking up these positions in the sorted data, for data set 8.3A the 1st quartile is 10 and the third quartile is 36, and for data set 8.3B the 1st quartile is 12 and the third quartile is 47.

2. The five-figure summaries are:

Table 9.11: Comparison of five-figure summaries for exam mark data

	Set A	Set B
Minimum	0	1
1st quartile	10	12
Median	18	27
3rd quartile	36	47
Maximum	47	50

3. Plotting the box plots on the same axis, we get something like this:

Figure 9.6: Box and whisker plot comparing two sets of students' multiple choice test marks

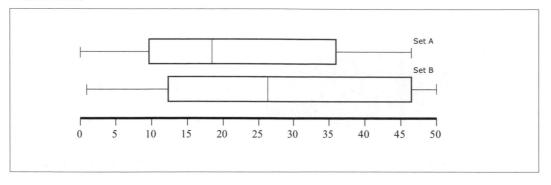

Activity 9.7: Finding the quartiles and IQR for a set of data

Using the formula:

$$\frac{n+1}{4} = \frac{30+1}{4} = \frac{31}{4} = 7.75$$

... the position of the lower quartile rounds to 8, so it's in the 8th position. That means its value is 18.

Using the formula:

$$3 \times \frac{n+1}{4} = 3 \times \frac{30+1}{4} = 3 \times \frac{31}{4} = 23.25$$

... the position of the upper quartile rounds to 23, so it's in the 23rd position. That means its value is 21.

The interquartile range is 21–18 = 3

This much more accurately describes how most of the data clusters around the median.

Activity 9.8: IQR 2

1. Using results from the previous chapter:
   - for set A the minimum is 0 and the maximum is 47, so the range is 47 – 0 = 47
   - for set B the minimum is 1 and the maximum is 50 so the range is 50 – 1 = 49

   Similarly, using results from activity 9.6:
   - for set A the 1st quartile is 10 and the 3rd quartile is 36, so the interquartile range is 36 – 10 = 26
   - for set B the 1st quartile is 12 and the 3rd quartile is 47, so the interquartile range is 47 – 12 = 35

2. Looking at the ranges, the two sets of data seem to give very similar results, since in both sets some students did very well while others did very badly.

The values for the interquartile range are lower than for the range, which is exactly what we would expect. They are still quite high, suggesting that the data is still quite spread out. The interquartile range for B is larger, which means that the middle half of this data set is more spread out.

Thus the interquartile range is saying that the marks in set B are more spread out, while the range can't see a difference. Because the range looks at all the data, it is affected by a few extreme values so it can't see what's happening in the middle. The interquartile range doesn't take the extreme values into account, so it can give a better impression of how the main central part of the data is spread.

Activity 9.9: Finding the variance and standard deviation for a set of data

1. The easiest way to start is by listing the data. You may find it helpful to rank it, but this is not necessary when you're working with the mean. Then fill in the columns. If you haven't already done so, you will need to find the mean – but we already did that in activity 8.7 of chapter 8.

Table 9.12: Finding the variance and standard deviation for the student ages data

	x	$x - \bar{x}$	$(x - \bar{x})^2$
	18	-2.27	5.14
	18	-2.27	5.14
	18	-2.27	5.14
	18	-2.27	5.14
	18	-2.27	5.14
	18	-2.27	5.14
	18	-2.27	5.14
	18	-2.27	5.14
	18	-2.27	5.14
	19	-1.27	1.60
	19	-1.27	1.60
	19	-1.27	1.60
	19	-1.27	1.60
	19	-1.27	1.60
	19	-1.27	1.60
	20	-0.27	0.07
	20	-0.27	0.07
	20	-0.27	0.07
	20	-0.27	0.07
	20	-0.27	0.07
	20	-0.27	0.07
	21	0.73	0.54
	21	0.73	0.54
	21	0.73	0.54
	21	0.73	0.54
	22	1.73	3.00
	22	1.73	3.00
	23	2.73	7.47
	25	4.73	22.40
	36	15.73	247.54
Σ	608		341.85

$$\bar{x} = \frac{\sum x}{n} = \frac{608}{30} = 20.27$$

$$s^2 = \frac{\sum(x-\overline{x})^2}{n} = \frac{341.87}{30} = 11.40$$

$$s = \sqrt{\frac{\sum(x-\overline{x})^2}{n}} = \sqrt{\frac{341.87}{30}} = 3.38$$

You may think this is a bit long-winded, especially as this data set has so many repeated numbers. You could have used the frequency distribution, just as you could have in chapter 8. But it is a method that will help you to work systematically and consistently.

2. Again, we need to know the mean of the data, but from activity 8.8 we already know that it's 10.875. We can set the calculations out in a table as usual to keep things clear.

Table 9.13 Finding the variance and standard deviation for the email data

	x	$x - \overline{x}$	$(x-\overline{x})^2$
	1	−9.875	97.52
	12	1.125	1.27
	22	11.125	123.77
	8	−2.875	8.27
	0	−10.875	118.27
	16	5.125	26.27
	24	13.125	172.27
	4	−6.875	47.27
	14	3.125	9.77
	3	−7.875	62.02
	1	−9.875	97.52
	11	0.125	0.02
	13	2.125	4.52
	29	18.125	328.52
	3	−7.875	62.02
	9	−1.875	3.52
	10	−0.875	0.77
	21	10.125	102.52
	19	8.125	66.02
	2	−8.875	78.77
	15	4.125	17.02
	1	−9.875	97.52
	16	5.125	26.27
	7	−3.875	15.02
Σ	261		1,566.73

Note that we don't need to sort the data into order for this calculation, since we're just adding up the columns.

So $$s^2 = \frac{\sum(x-\overline{x})^2}{n} = \frac{1566.73}{24} = 65.28$$

Then $$s = \sqrt{\frac{\sum(x-\overline{x})^2}{n}} = \sqrt{65.28} = 8.08$$

So the variance is 65.28, and the standard deviation is 8.08.

Activity 9.10: Finding the coefficient of variation for a set of data

1. The mean for the student age data was 20.27 and the standard deviation was 3.38.
   So the coefficient of variation is 3.38 ÷ 20.27 x 100= 16.67.

2. The mean for the email data was 10.875 and the standard deviation was 8.08.
   So the coefficient of variation is 8.08 ÷ 10.875 × 100 = 74.30.

CHAPTER 10

Correlation analysis

OVERVIEW

In previous chapters we have looked at ways of summarising and describing the information contained in a single variable (or set of data). Specifically in chapter 7 we looked at some tabular and graphical methods which can be used to do this. In chapters 8 and 9 we looked at some numerical ways of summarising data. In this chapter we will turn our attention to methods which will enable us to analyse two variables with a view to examining to what extent they are related. Correlation will be introduced as a general measure of the strength of association between two numerical variables that are in some way related.

The chapter begins with the exploration of a simple example. You will be guided through the use of graphical methods that are used in exploring relationships between numerical variables and you will then be shown how to calculate a correlation coefficient by hand. You will learn when to use a correlation coefficient sensibly and how to interpret it correctly. The chapter will conclude with giving you ideas of how this topic extends to other areas which are not formally covered in this module.

Learning outcomes On completion of this chapter, you should be able to:

- Construct and use a scatter plot to examine the relationship between two variables

- Explain what is meant by a linear relationship between two variables

- Describe the term 'correlation coefficient'

- Calculate and interpret a correlation coefficient

- Discuss the problems of spurious correlation.

10.1 Relationships between two numerical variables

When the values in one variable (data set) are related to the values in another variable, the two variables are said to be *correlated*. The basic idea of correlation analysis is to assess the strength of an association that may exist between two variables. It is fairly easy to come up with examples of variables that you would expect to be correlated. For example:

- The distance of a journey and the time it takes to make it
- The age and height of a child
- The price of a product and the demand for (or number of units of) the product sold
- The time a student spends studying and the mark they achieve on the examination.

Example 10.1

Suppose the sales manager of a national photocopier supply company wants to determine if there is a relationship between the number of sales calls made by its sales representatives in a month and the number of photocopiers sold that month. The data in table 10.1 shows the number of sales calls made by 10 randomly selected representatives last month and the number of copiers sold.

Table 10.1: Sales of photocopiers

Number of calls made by the sales representative	20	40	20	30	10	10	20	20	20	30
Number of photocopiers sold	30	60	40	60	30	40	40	50	30	70

Source of data: Lind, Mason and Marchal,
Basic Statistics for Business and Economics,
Third edition, McGraw-Hill (page 360)

There are two particular questions we want to answer when completing a correlation analysis on data like this:

- Is there any relationship between these two variables? In other words, is there a relationship between the number of calls made by a sales representative and the number of copiers sold?
- How strong is this relationship, if it exists?

Scatter plots

The first step in answering both of these questions is to produce a *scatter plot* of the data. This is just a plot of the data, ten pairs of observations in this instance, with one variable on the *x*-axis and the other variable on the *y*-axis. A plot carefully drawn by hand of the data in table 10.1 might look like that in figure 10.1.

To produce this plot by hand, we begin with the first pair of observations. This relates to a representative who made 20 sales calls last month and sold 30 photocopiers, so $x = 20$ and $y = 30$. To plot this point, move along the horizontal axis to $x = 20$, then go vertically to $y = 30$ and put a dot at the intersection. This is repeated until each pair of observations has been plotted.

We could have chosen to put the number of photocopiers sold on the *x*-axis and the number of sales calls made on the *y*-axis. In correlation analysis it does not matter which variable you choose to plot on each axis.

Figure 10.1: Scatter plot of number of photocopiers sold against number of sales calls made

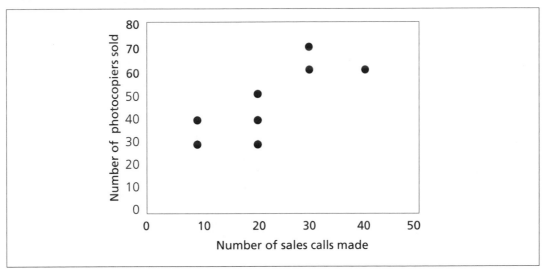

This plot does seem to suggest that there is a relationship between the number of sales calls made and the number of photocopiers sold, as the points are tending to go up from the bottom left to the top right. The representatives who make more sales calls seem to sell more copiers. The relationship also appears to be linear: that is, it could reasonably be described by putting a straight line through the points as shown in figure 10.2. The points do not lie perfectly on the straight line, but this is not a problem. The straight line does a reasonable job of describing the general pattern of the points and hence the relationship is linear.

Figure 10.2: Scatter plot showing linear relationship

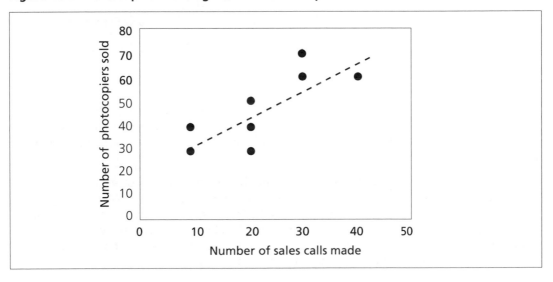

Activity 10.1

The relationship between the cost of milk and sales 1

A milkman observes the selling price and sales volume of milk (in gallons) for eight randomly selected weeks. Table 10.2 details the data he has collected.

Produce a scatter plot of this data. Does there appear to be a relationship between these two variables and, if so, is it linear? How would you describe this relationship?

Table 10.2: Selling price and volume of sales of milk

Weekly sales (gallons)	10	5	12	10	15	7	12	17
Selling price per pint (in pence)	33	37	35	36	31	36	34	30

Activity 10.2

The relationship between computer speed and runtime

A computer magazine tests a number of computers by seeing how long they take to run a standard test program. The results are shown in table 10.3.

Produce a scatter plot of this data. Does there appear to be a relationship between these variables and, if so, is it linear?

Table 10.3: Clock speed and time to run test program

Clock speed (MHz)	1.5	3.0	2.8	2.1	1.6	2.4	3.1	1.0	2.7	3.0
Run-time (seconds)	85	39	47	63	65	43	48	118	48	35

Source of data: hypothetical

It's important to remember that all we're doing at the moment is investigating whether there seems to be a relationship between two variables. We might notice that high values of one variable seem to be associated with high values of the other, but we're not saying that changes in one variable *cause* changes in the other.

For example, you may think that one sales representative sells more photocopiers than another *because* the first one makes more sales calls, or that the milkman sells more milk one week than another *because* he charged a lower price in the first week. This might well be true, but the point is we haven't proved it just by drawing a scatter plot. All the scatter plot does is tell us that there seems to be a relationship that's worth investigating. It doesn't prove that the relationship actually exists. We'll come back to this idea later on.

10.2 Degrees of correlation

So scatter plots can help us examine the data visually to see if a relationship exists between two variables. In so doing we can decide whether the relationship can be described by a straight line; in other words, whether it is linear. If it is, the scatter plot can help us decide whether the line which describes the relationship slopes up (positive slope) or slopes down (negative slope). This is one of the first important characteristics of correlation: whether it is a positive or a negative relationship.

Positive correlation

As one variable increases, so does the other. Low values of one variable are associated with low values of the other variable and high values of one variable are associated with high values of the other. This is evident on a scatter plot by the suggestion of an upward sloping straight line that could be used to describe the data. This was the situation we had in the example on photocopiers. The line in figure 10.2 slopes upwards. Interpreting this in the context of the data leads us to conclude that *as the number of sales calls increases, the number of photocopiers sold also increases*.

Negative correlation

As one variable increases, the other decreases. So high values of one variable are associated with low values of the other variable. This is evident on the scatter plot by the suggestion of a downward sloping straight line that could be used to describe the data. This was the situation we had in the activity on milk sales, where the data slopes downwards. Interpreting this in the context of the data leads us to conclude that *as the price of milk goes up (increases), the sales go down (decrease).*

So the positive or negative nature of the relationship between the two variables is one of the things which needs to be explained in the context of the data. The other thing that needs to be considered is how strong the relationship between the two variables actually is. Are increases in one variable matched quite closely by increases in the other, or is the relationship a bit weaker than that? This can be seen on the scatter plot by judging how closely the points cluster around the line that describes the relationship. Do the points appear to be following a suggested straight line very closely or are they more spread out about the suggested straight line? These differences or degrees in the strength of the correlation are best considered by looking at some plots.

Perfect correlation

If the relationship or correlation between the two variables is perfect, the points in the scatter plot will lie exactly on a straight line: an exact linear relationship exists between the two variables. If this relationship is positive the line will slope upward, and if it is negative the line will slope downward, as shown in figure 10.3.

Figure 10.3: Perfect correlation scatter plots

Partial correlation

In practice, the relationship between two variables is rarely perfect. This is evident in a scatter plot when the points look as if they are following the pattern of a straight line but they are not located perfectly on that straight line. If the points are quite close to the suggested straight line then the relationship, although not perfect, can be considered to be quite strong. As the points get more spread about the suggested straight line the relationship gets weaker. The following two plots indicate what you can expect to see on a scatter plot. These plots demonstrate partial correlation, which is what you are most likely to meet in practice. In the first plot, in figure 10.4, the points seem to follow a straight-line pattern very closely, so this is strong partial correlation. In contrast, in the second plot, the points are more spread out, which is why the plot has been labelled as a medium-strength relationship.

Figure 10.4: Partial correlation scatter plots

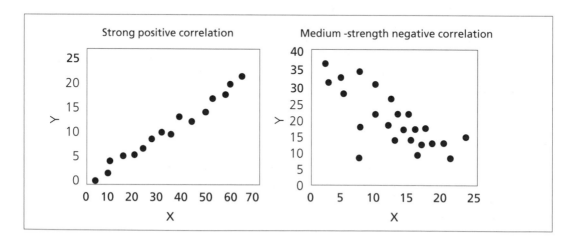

No correlation

There is no correlation (or relationship) between the two variables if the scatter plot displays a random scatter of points which are neither tending to go up or down. This means that a linear relationship does not exist between these two variables.

Figure 10.5: No correlation scatter plots

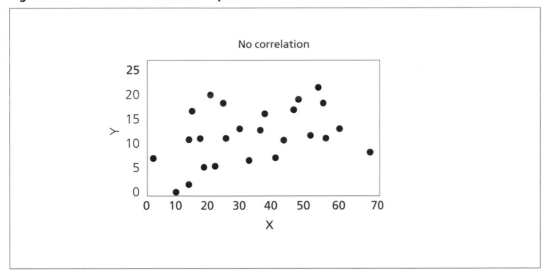

Activity 10.3

The relationship between the cost of milk and sales 2

Refer back to the scatter plot you produced for activity 10.1. Does this graph show perfect or partial correlation? If it is partial correlation, how strong do you think the relationship is?

Do the same for the scatter plot for activity 10.2.

10.3 Product moment correlation coefficient

We now know how to use scatter plots to examine if a relationship exists between two variables and, if it does, whether it is positive or negative in nature. Using the scatter plot we can also make a judgement about how strong the relationship is. The product moment correlation coefficient, which we will call r, is a numerical method of doing almost the same thing, and gives a statistical summary of the relationship between two variables. If we calculate r, we will know how strong a relationship is between two variables, and whether it is positive or negative in nature.

The product moment correlation coefficient r measures the strength of a *linear* relationship. Two variables could be very strongly related, but if a straight line cannot describe this relationship it is not appropriate to use r as a statistical summary measure. For this reason it is important that you plot the data first to check that the relationship is linear before proceeding to calculate r.

The value of r will always be between –1 and +1. If you get a value outside this range you have made a mistake. The sign of r tells us if the relationship between the two variables is positive or negative. The value of r tells us if the relationship is strong or weak, as the following explains.

$r = +1$:   Perfect positive correlation

$0 < r < 1$: Partial positive correlation. The closer r gets to 1, the stronger the relationship

$r = 0$:   No correlation

$-1 < r < 0$: Partial negative correlation. The closer r gets to -1, the stronger the relationship

$r = -1$:   Perfect negative correlation

If we translate this information onto the plots we used to explain degrees of correlation, we get the likely values for r as shown in figure 10.6.

Figure 10.6: Likely values for r

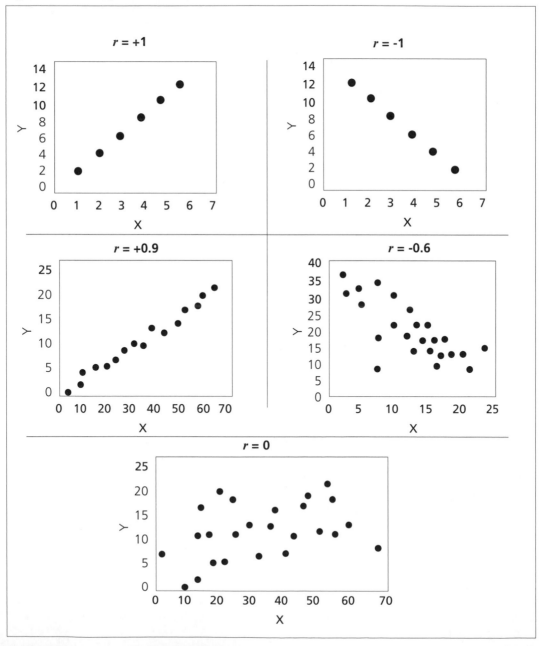

10.4 Calculating a product moment correlation coefficient

The product moment correlation coefficient is calculated using the following formula:

Formula 10.1: Product moment correlation coefficient, *r*

$$r = \frac{\sum (x - \bar{x})(y - \bar{y})}{\sqrt{\sum (x - \bar{x})^2 \times \sum (y - \bar{y})^2}}$$

In this formula the letters x and y represent the pairs of data for the two variables. Also:

$\bar{x} = \frac{\sum x}{n}$ is the mean of the variable x as explained in earlier chapters

$\bar{y} = \frac{\sum y}{n}$ is the mean of the variable y

The formula looks quite complicated, but if you are quite methodical in the way you approach it you will find it is not that hard to evaluate. Remember that we have two variables so we shall refer to one as x and the other as y.

Example 10.2

For this example we will return to the photocopier data we considered in example 10.1. The data is reproduced below.

Table 10.1: Sales of photocopiers

Number of calls made by the sales representative	20	40	20	30	10	10	20	20	20	30
Number of photocopiers sold	30	60	40	60	30	40	40	50	30	70

Source of data: Lind, Mason and Marchal,
Basic Statistics for Business and Economics,
Third edition, McGraw-Hill (page 360)

We already know, from the scatter plot in figure 10.2, that this data follows a linear relationship as the points tend to follow a straight line, so it would be sensible to use the product moment correlation coefficient to summarise this relationship. We also know, from figure 10.2, that the relationship is positive as the scatter plot slopes upward, so we would expect r to be positive for this data.

Producing a table of calculations is the best way to evaluate r. First we need to decide which variable will be represented by x and which will be called y. For correlation analysis it does not matter which variable is called x and y. So, for our table of calculations we will refer to the number of calls as x and the number of sales as y. The data can then be written down as shown in table 10.4 ready for calculating r.

Table 10.4: Preparing to calculate r

	No. of calls, x	No. of sales, y					
	20	30					
	40	60					
	20	40					
	30	60					
	10	30					
	10	40					
	20	40					
	20	50					
	20	30					
	30	70					
Total							

The empty columns and last row of this table will get filled in as we work our way through the calculation of r.

The formula for r requires us to use the mean for variables x and y, so we will calculate these first.

If we add up the values in the first column relating to variable x, (number of calls), we get:

Sum of variable $x =$

Σx= 20+40+20+30+10+10+20+20+20+30 = 220, and this has been entered into the totals row in table 10.5.

As there are 10 salesman in the study we know that n = 10 and from here we can calculate the mean to be:

$$\bar{x} = \frac{\Sigma x}{n} = \frac{220}{10} = 22$$

We can do the same with the second variable y (number of sales), to give:

Sum of variable $y =$

Σy =30+60+40+60+30+40+40+50+30+70=450 which has been entered into the totals row in table 10.5.

Then $\bar{y} = \frac{\Sigma y}{n} = \frac{450}{10} = 45$

The next step is to subtract the mean $\bar{x}$ from each observation of the variable x. The results are shown in the column headed $(x-\bar{x})$ in table 10.5. This is then repeated for variable y and the results are shown in the column headed $(y-\bar{y})$ in table 10.5.

The values of $(x-\bar{x})$ are then squared to give the results in the column headed $(x-\bar{x})^2$ of table 10.5 and the values of $(y-\bar{y})$ are squared to give the results in the column headed $(y-\bar{y})^2$ of table 10.5. The last column to create is headed $(x-\bar{x}) \times (y-\bar{y})$ which is created by taking each value in the $(x-\bar{x})$ column and multiplying it by the corresponding value (value on the same row) of the $(y-\bar{y})$ column. All of these calculations are presented in table 10.5. You can see that some of the calculations are similar to those we did when calculating variances in chapter 9. Take some time looking at this table to make sure you understand how to create each additional column.

Table 10.5: Evaluating r

x	y	$x-\bar{x}$	$y-\bar{y}$	$(x-\bar{x})^2$	$(y-\bar{y})^2$	$(x-\bar{x}) \times (y-\bar{y})$
20	30	20-22 = -2	30-45 = -15	-2^2 = 4	-15^2 = 225	-2×-15 = 30
40	60	40-22 = 18	60-45 = 15	18^2 = 324	15^2 = 225	18×15 = 270
20	40	20-22 = -2	40-45 = -5	-2^2 = 4	-5^2 = 25	-2×-5 = 10
30	60	30-22 = 8	60-45 = 15	8^2 = 64	15^2 = 225	8×15 = 120
10	30	10-22 = -12	30-45 = -15	-12^2 = 144	-15^2 = 225	-12×-15 = 180
10	40	10-22 = -12	40-45 = -5	-12^2 = 144	-5^2 = 25	-12×-5 = 60
20	40	20-22 = -2	40-45 = -5	-2^2 = 4	-5^2 = 25	-2×-5 = 10
20	50	20-22 = -2	50-45 = 5	-2^2 = 4	5^2 = 25	-2×5 = -10
20	30	20-22 = -2	30-45 = -15	-2^2 = 4	-15^2 = 225	-2×-15 = 30
30	70	30-22 = 8	70-45 = 25	8^2 = 64	25^2 = 625	8×25 = 200
Σx = 220	Σy = 450			$\Sigma(x-\bar{x})^2$ = 760	$\Sigma(y-\bar{y})^2$ = 1850	$\Sigma(x-\bar{x}) \times (y-\bar{y})$ = 900

Summing the values in the columns headed $(x-\bar{x})^2$, $(y-\bar{y})^2$ and $(x-\bar{x}) \times (y-\bar{y})$ gives the results below, which are also presented in the totals row of table 10.5:

$\Sigma(x-\bar{x})^2 = 4+324+4+64+144+144+4+4+4+64 = 760$

$\Sigma(y-\bar{y})^2 = 225+225+25+225+225+25+25+25+225+625 = 1,850$

$\Sigma(x-\bar{x}) \times (y-\bar{y}) = 30+270+10+120+180+60+10+(-10)+30+200 = 900.$

We now need to substitute these three summed terms into the formula for the product moment correlation coefficient, which is:

$$r = \frac{\sum(x-\bar{x}) \times (y-\bar{y})}{\sqrt{\sum(x-\bar{x})^2 \times (y-\bar{y})^2}}$$

to give

$$r = \frac{900}{\sqrt{760 \times 1850}} = \frac{900}{\sqrt{1406000}} = \frac{900}{\sqrt{1185.75}} = 0.7590$$

The value of the product moment correlation coefficient for this data is 0.7590. It is a positive value, as expected from the scatter plot, and suggests that if the sales representatives make more calls then they will make more sales of photocopiers. The value of the correlation coefficient is less than 1 so this is not a perfect relationship, but $r = 0.75$ is quite close to 1, so the relationship is reasonably strong.

Activity 10.4

The relationship between the cost of milk and sales 3

Refer back to the data you worked on in activities 10.1 and 10.3. Calculate the product moment correlation coefficient r and interpret the result.

Activity 10.5

Relationship between computer speed and runtime 2

Refer back to the computer speed data you worked on in activity 10.2. Calculate the product moment correlation coefficient r and interpret the result.

10.5 Correlation and causation

There has been little mention throughout this chapter about the notion of cause or effect. What we mean by this is that we have concentrated on exploring if a relationship exists between two variables but we have avoided saying that movement in one variable is causing changes in another. Let us go back to the first example where we were investigating the number of calls made by a salesman in a given month and his corresponding level of sales. All we tried to do in this example was investigate if there is a relationship between these two variables. Intuitively, it would seem sensible to say that if the salesman makes more calls he will generate more sales. This now moves into the area of cause and effect: we are now saying that the number of calls the salesman makes *explains* his level of sales. Another way you may see this expressed in some textbooks is that the level of sales *responds* to the number of calls made. In some more advanced analyses, such as regression analysis, it is important that we decide which variable is doing the explaining before we start any calculations. Once we have done this, we call one variable the *response* and the other the *explanatory*. For example 10.1 these would be:

Response variable

Number of sales: the number of sales *responds to* the number of calls made.

Explanatory variable

Number of calls: the number of calls made *explains* the level of sales achieved.

None of this matters in calculating a correlation coefficient. We do not need to specify which is the response variable when calculating r because the correlation coefficient just measures linear association and has nothing directly to say about cause and effect. However, it is quite easy to forget this when interpreting r and draw conclusions about cause and effect that are inaccurate. Two variables may be very strongly correlated when there is no direct cause and effect relationship between them. This is called *spurious correlation*. This sort of situation can arise when the two variables used in the correlation analysis are both affected by a third variable which was not included in the initial analysis.

Example 10.3

The following scatter plot in figure 10.7 of the total federal debt versus the number of golfers in America shows a strong correlation. Would cutting the number of golfers, by taxing them or making the sport illegal, reduce the federal debt?

Figure 10.7: Scatter plot of federal debt versus number of golfers in America

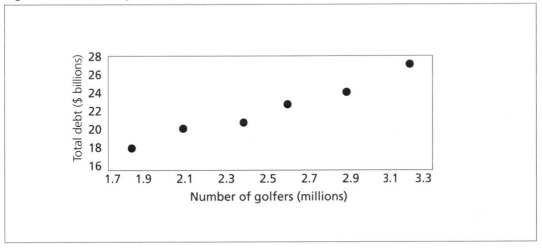

Based on data presented in Johnson and Wichern, *Business Statistics: Decision making with data*.
Wiley Press

Clearly, from the plot, the correlation is strong and positive, implying that as the number of golfers increase, the size of the debt in America also increases. Commonsense should make us think this conclusion cannot be correct and there is no cause and effect between the federal debt and the number of golfers. This looks like spurious correlation. The scatter plot is repeated in figure 10.8, but each point has been labelled according to the year, with '96' standing for 1996, etc. Many things can change over the course of time and the year is just a substitute, or proxy, for all of them. The years go up the points from left to right in the scatter plot. Clearly, both of the variables have increased over time, for all sorts of reasons. It is this common increase over time that makes the two variables strongly correlated – but there is no cause and effect link between them.

Figure 10.8: Scatter plot of federal debt versus number of golfers in America indicating the year

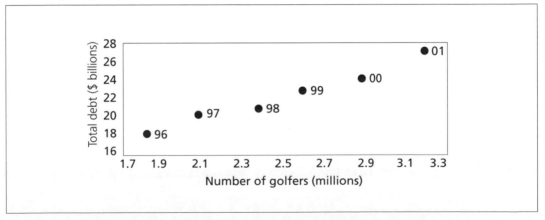

Activity 10.6

Does watching television extend your life?

Moore, McCabe, Duckworth and Sclove report in *The Practice of Business Statistics* that there is a strong positive correlation between the two variables x and y where x is the number of television sets per person in a country and y is the average life expectancy for that country. Can you conclude that we could lengthen the lives of people in a country with poor life expectancy by shipping them TV sets?

Activity 10.7

Do policemen affect the price of chocolate?

There is a strong positive correlation between the two variables x and y where x is the average wage of a policeman in the UK and y is the average price of a chocolate bar in the UK. Can you conclude that policemen are being paid with the earnings of chocolate companies? Can you conclude that policemen spend a lot of their wages on chocolate, so that when police wages rise the policemen have more money to spend on chocolate so the price of chocolate goes up as well?

Activity 10.8

Do bananas cure polio?

There is a strong negative correlation between the two variables x and y where x is the number of bananas eaten in the UK and y is the number of people suffering from polio in the UK. Can we conclude that bananas cure polio?

10.6 Summary

In this chapter we saw how to measure a possible connection between two sets of data, using correlation. We looked at when this idea is appropriate, and when it is not. Correlation is an important technique, since there are many situations in which we want to find out whether one quantity might have an effect on another.

10.7 Review questions

Refer to the scatter plots below to answer review questions 10.1–10.6.

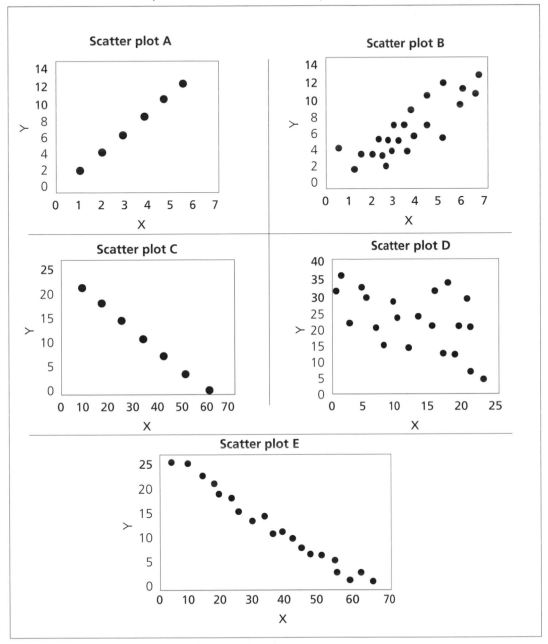

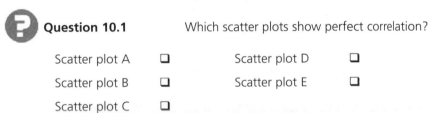

Question 10.1   Which scatter plots show perfect correlation?

Scatter plot A	❏	
Scatter plot B	❏	
Scatter plot C	❏	
Scatter plot D	❏	
Scatter plot E	❏	

Question 10.2 Which scatter plot shows no correlation?

Scatter plot A ❏   Scatter plot D ❏

Scatter plot B ❏   Scatter plot E ❏

Scatter plot C ❏

Question 10.3 Which scatter plots show a negative relationship?

Scatter plot A ❏   Scatter plot D ❏

Scatter plot B ❏   Scatter plot E ❏

Scatter plot C ❏

Question 10.4 Of those scatter plots showing partial correlation, which has the strongest relationship?

Scatter plot A ❏   Scatter plot D ❏

Scatter plot B ❏   Scatter plot E ❏

Scatter plot C ❏

Question 10.5

What is the correlation coefficient of scatter plot D?

Question 10.6

Which scatter plot would have a correlation coefficient of –1?

Question 10.7 Assessing relationships between numerical variables using scatter plots

For each of the data sets below produce a scatter plot. Does there appear to be a relationship between the pairs of variables in each data set? If so, what is this relationship and how would you describe it?

Data set 10.1

Price ($)	3	4	6	7	8	10
Sales (in hundreds)	12	6	11	3	7	1

Data set 10.2

x		10	11	12	13	14	15	16	17
y		8	13	5	11	7	4	9	11

Data set 10.3

Does fast driving waste fuel? How does the fuel consumption of a car change as its speed increases? The data below relate to a British Ford Escort car. Speed is measured in kilometres per hour and fuel consumption is measured in litres of petrol used per 100 kilometres travelled.

Speed (km/hr)	Fuel used (litres per 100km)	Speed (km/hr)	Fuel used (litres per 100km)
10	21.00	90	7.57
20	13.00	100	8.27
30	10.00	110	9.03
40	8.00	120	9.87
50	7.00	130	10.79
60	5.90	140	11.77
70	6.30	150	12.83
80	6.95		

Data source: Moore, McCabe, Duckworth and Sclove
The Practice of Business Statistics, Freeman Press (page 93)

Data set 10.4

A food industry group asked 3,368 people to guess the number of calories in each of several common foods. Table 10.6 displays the averages of their guesses and the correct number of calories.

Table 10.6: Comparison of guessed and correct calories for 10 different foods

Food	Guessed calories	Correct calories
8oz of full-fat milk	196	159
5oz spaghetti with tomato sauce	394	163
5oz macaroni with cheese	350	269
One slice wholewheat bread	117	61
One slice white bread	136	76
2oz candy bar	364	260
Saltine cracker	74	12
Medium-size apple	107	80
Medium-size potato	160	88
Cream-filled cake	419	160

Data source: Moore, McCabe, Duckworth and Sclove
The Practice of Business Statistics, Freeman Press (page 109)

Question 10.8: Assessing perfect and partial correlation

Look back at the plots you produced in review question 10.7. Would you say these data sets display perfect, partial or no correlation? If the correlation is partial, how strong do you think the relationship is?

 Question 10.9: Calculating a product moment correlation coefficient

Data sets 10.1 and 10.2

Given your work in review questions 10.7 and 10.8, what would you expect the value of the product moment correlation coefficient to be for these data sets? Evaluate r for each data set and interpret its value.

Data sets 10.3 and 10.4

Given your work in review questions 10.7 and 10.8, what would you expect the value of the product moment correlation coefficient to be for these data sets? Evaluate r for each data set and interpret its value. You make use of the following quantities for each data set in your evaluation:

Data set 10.3

In the following sums, x relates to speed and y relates to fuel consumption.

$\Sigma(x-\bar{x})^2 = 28.000$ $\Sigma(y-\bar{y})^2 = 204.227$ $\Sigma(x-\bar{x}) \times (y-\bar{y}) = 410.4$

Data set 10.3

In the following sums, x relates to the guessed calories and y relates to the correct calories.

$\Sigma(x-\bar{x})^2 = 162,070$ $\Sigma(y-\bar{y})^2 = 64,837.6$ $\Sigma(x-\bar{x}) \times (y-\bar{y}) = 84,519.4$

 Question 10.10: Assessing spurious correlation

The following data details the number of 18-hole golf courses in America for 12 consecutive years, and the corresponding number of divorces in the same year.

Table 10.7: Spurious correlation?

Year	Number of 18-hole golf courses	Number of divorces in America (in millions)
1	2,725	2.9
2	3,769	3.5
3	4,845	4.3
4	6,282	6.5
5	6,551	8.0
6	6,699	8.6
7	6,787	8.8
8	6,856	9.9
9	6,944	10.8
10	7,059	11.5
11	7,125	11.6
12	7,230	12.3

Data source: McClave and Benson, *Statistics for Business and Economics* 4th edition (page 521)

It can be shown that the product moment correlation coefficient is 0.917, which is very close to a perfect positive correlation. Can we conclude that people are spending too long on the golf course, which is causing their marriages to break down and hence resulting in divorce?

10.8 Answers to review questions

Question 10.1

Scatter plots A and C.

Feedback: Perfect correlation is indicated on a scatter plot by the points following a straight line exactly. This only happens in plots A and C.

Question 10.2

Scatter plot D.

Feedback: No correlation means there is no linear relationship between x and y evident on the scatter plot. This is only the case for scatter plot D.

Question 10.3

Scatter plots C and E.

Feedback: A downward sloping line in a scatter plot indicates a negative relationship. This is only true for plots C and E.

Question 10.4

Scatter plot E.

Feedback: The points following the general pattern of a straight line without being exactly on that straight line indicate a partial relationship on a scatter plot. The closer the points are to the suggested line, the stronger the relationship is. Scatter plots B and E suggest a partial relationship. The points are closer to the suggested straight line in scatter plot E, therefore this is the strongest partial relationship.

Question 10.5

0

Feedback: The points suggest no relationship between x and y. This would result in a correlation coefficient of 0.

Question 10.6

Scatter plot C.

Feedback: If the correlation coefficient is −1, this indicates a perfect negative relationship, which corresponds to plot C.

Question 10.7:

Data set 10.1

The scatter plot is shown in figure 10.9. There is a relationship between these two data sets as the points could be described as following a downward sloping line. Therefore this relationship is linear. As the line slopes downward, the nature of this relationship is that as price (x) increases, sales (y) decrease.

Figure 10.9: Scatter plot for data set 10.1

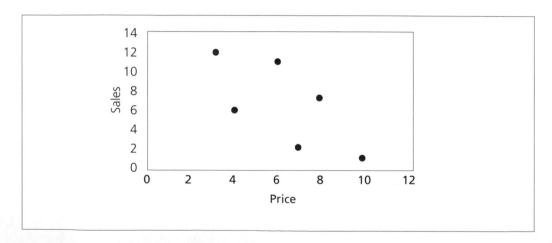

Data set 10.2

The scatter plot is shown in figure 10.10. The pattern of the points in this plot is completely random and does not appear to be sloping either upward or downward. Therefore there is no relationship between these two variables.

Figure 10.10: Scatter plot for data set 10.2

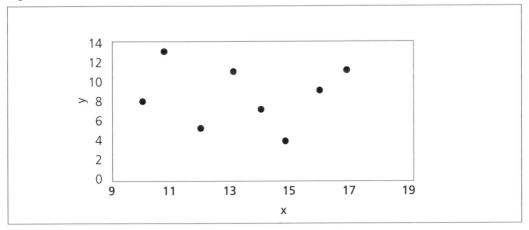

Data set 10.3

The scatter plot is shown in figure 10.11. There is clearly a relationship between these two data sets but a straight line cannot describe it. The points follow a curved pattern very closely. Therefore this relationship is not linear. At low speeds, fuel consumption is quite high – but the fuel consumption drops as speed increases to about 60km/hr. As speed increases beyond this, the fuel consumption starts to rise again.

Figure 10.11: Scatter plot for data set 10.3

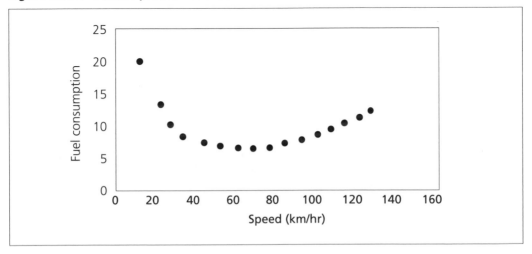

Data set 10.4

The scatter plot is shown in figure 10.12. There is a relationship between these two data sets as the points could be described as following an upward sloping line. Therefore this relationship is linear. As the line slopes upward, the nature of this relationship is that as guessed calories (x) increases, correct calories (y) decreases. Therefore, people's perceptions of what foods have a lot of calories seem to be correct.

Figure 10.12: Scatter plot for data set 10.4

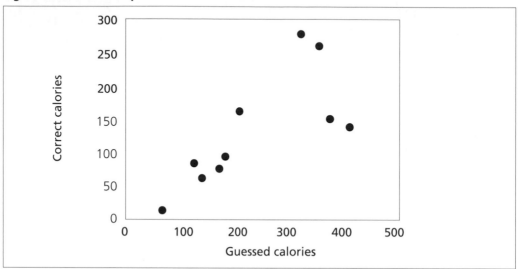

Question 10.8: Assessing perfect and partial correlation

Data set 10.1

The points are not lying exactly on a straight line so this is *partial*, rather than *perfect*, correlation. The points are quite spread out about the suggested line, so this is not a very strong relationship.

Data set 10.2

There is no relationship between the two variables in this plot. Therefore there is no correlation at all.

Data set 10.3

The points are following a very good relationship in this plot but it is not a straight line. It is hard to comment on whether the relationship is perfect or partial as it is not a straight line. Without knowing what the relationship is, we cannot make a judgement about whether the points are following the relationship, exactly or partially. The pattern of the points is quite clear, so the relationship, whatever it is, is quite strong.

Data set 10.4

The points are not lying exactly on a straight line, so this is partial, rather than perfect, correlation. The points are reasonably tightly clustered around the suggested line, with the exception of two points on the far right which look a bit different. If we ignore those two points for this discussion, we would say this relationship is quite strong.

Question 10.9: Calculating a product moment correlation coefficient

Data set 10.1

From the work in review questions 10.7 and 10.8 we would expect the correlation coefficient to be negative and not that strong.

Table 10.8 gives the details of the calculations for r. Variable x refers to the price and y to the sales.

There are six observations in each variable so $n = 6$:

$$\bar{x} = \frac{\sum x}{n} = \frac{38}{6} = 6.3 \text{ and } \bar{y} = \frac{\sum y}{n} = \frac{40}{6} = 6.7$$

N.B. These are the values of $\bar{x}$ and $\bar{y}$ correct to 1 decimal place. If you have used more than 1 decimal place in your solution, there may be slight differences between your answers and the calculation provided on the other page. However, your final answer for r should be the same to the first decimal place as the one overleaf.

Table 10.8: Evaluating r for data set 10.1

x	y	$x-\bar{x}$	$y-\bar{y}$	$(x-\bar{x})^2$	$(y-\bar{y})^2$	$(x-\bar{x}) \times (y-\bar{y})$
3	12	-3.3	5.3	10.89	28.09	-17.49
4	6	-2.3	-0.7	5.29	0.49	1.61
6	11	-0.3	4.3	0.09	18.49	-1.29
7	3	0.7	-3.7	0.49	13.69	-2.59
8	7	1.7	0.3	2.89	0.09	0.51
10	1	3.7	-5.7	13.69	32.49	-21.09
$\sum x$ = 38	$\sum y$ = 40			$\sum(x-\bar{x})^2$ = 33.34	$\sum(y-\bar{y})^2$ = 93.34	$\sum(x-\bar{x}) \times (y-\bar{y})$ = –40.34

Substituting the relevant sums into the formula:

$$r = \frac{\sum(x-\bar{x})(y-\bar{y})}{\sqrt{\sum(x-\bar{x})^2 \times \sum(y-\bar{y})^2}}$$

gives:

$$r = \frac{-40.34}{\sqrt{33.34 \times 93.34}} = \frac{-40.34}{\sqrt{3111.9556}} = \frac{-40.34}{\sqrt{55.78}} = -0.72$$

This correlation coefficient is negative, as expected, so as the price goes up, the amount sold goes down. This is not a perfect relationship as the correlation coefficient is not equal to –1, and with a value of –0.72 it is indicating a reasonably strong relationship.

Data set 10.2

From the work in review questions 10.7 and 10.8 we have determined that there is no relationship between these two variables, so we would expect r to be approximately zero.

Table 10.9 gives the details of the calculations for r. There are eight observations in each variable so $n = 8$.

$$\bar{x} = \frac{\sum x}{n} = \frac{108}{8} = 13.5 \text{ and } \bar{y} = \frac{\sum y}{n} = \frac{68}{8} = 8.5$$

Substituting the relevant sums from table 10.9 into the formula

$$r = \frac{\sum(x-\bar{x})(y-\bar{y})}{\sqrt{\sum(x-\bar{x})^2 \times \sum(y-\bar{y})^2}}$$

gives

$$r = \frac{-3}{\sqrt{42 \times 68}} = \frac{-3}{\sqrt{2856}} = \frac{-3}{\sqrt{53.44}} = -0.056$$

This correlation coefficient is very close to zero as expected, so there is almost no relationship between the two variables in data set 10.2.

Table 10.9: Evaluating r for data set 10.2

x	y	$x-\bar{x}$	$y-\bar{y}$	$(x-\bar{x})^2$	$(y-\bar{y})^2$	$(x-\bar{x}) \times (y-\bar{y})$
10	8	-3.5	-0.5	12.25	0.25	1.75
11	13	-3.5	4.5	6.25	20.25	-11.25
12	5	-1.5	-3.5	2.25	12.25	5.25
13	11	-0.5	2.5	0.25	6.25	-1.25
14	7	0.5	-1.5	0.25	2.25	-0.75
15	4	1.5	-4.5	2.25	12.25	-6.75
16	9	2.5	0.5	6.25	0.25	1.25
17	11	3.5	2.5	12.25	6.25	8.75
$\sum x$ = 108	$\sum y$ = 68			$\sum(x-\bar{x})^2$ = 42	$\sum(y-\bar{y})^2$ = 68	$\sum(x-\bar{x}) \times (y-\bar{y})$ = -3

Data set 10.3

We are not really sure what to expect here for r. The two variables in data set 10.3 are clearly strongly related but the relationship is not linear. As r measures the strength of a linear relationship, it is inappropriate to use it for this data set. However, if we do continue to evaluate r we get the following:

Substituting the relevant sums from the question into the formula:

$$r = \frac{\sum(x-\bar{x})(y-\bar{y})}{\sqrt{\sum(x-\bar{x})^2 \times \sum(y-\bar{y})^2}}$$

gives

$$r = \frac{-410.4}{\sqrt{28000 \times 204.227}} = \frac{-410.4}{\sqrt{2391.31}} = -0.17$$

As this value is not that far from zero, it is suggesting a weak relationship exists between the two variables. The scatter plot in review question 10.7 indicates that this is not the case. The two variables are strongly related but not in a linear way, hence the low value for r as the product moment correlation coefficient measures linear association.

Data set 10.4

From the work in review questions 10.1 and 10.2 we would expect the correlation coefficient to be positive and quite strong.

Substituting the relevant sums from the question into the formula:

$$r = \frac{\sum (x - \bar{x})(y - \bar{y})}{\sqrt{\sum (x - \bar{x})^2 \times \sum (y - \bar{y})^2}}$$

gives:

$$r = \frac{84519.4}{\sqrt{162070 \times 64837.6}} = \frac{84519.4}{\sqrt{102509.66}} = 0.82$$

This is positive and quite strong as expected.

Question 10.10: Assessing spurious correlation

Common sense should be alerting you to the possibility that this is an example of spurious correlation. If you read down the column detailing the number of golf courses, you can see that this figure has consistently increased from one year to the next during the 12 years relating to this study. This increasing trend is also apparent in the column detailing the number of divorces. Both of these variables are increasing over time, which causes the spurious strong positive correlation between the two variables. It would be incorrect to infer that the increasing number of golf courses is causing the increase in the number of divorces.

10.9 Feedback on activities

Activity 10.1: The relationship between cost of milk and sales

Figure 10.13 (as seen earlier) gives you an idea of what your scatter plot should look like. You may have decided to put selling price on the *y*-axis instead of weekly sales. This will make your scatter plot look a little different but it does not matter. Both graphs are equally correct. As you can see from figure 10.13 there does appear to be a relationship between these two data sets. You could easily draw a straight line through the points in figure 10.13 and it would do a good job of describing the general pattern of the points, therefore the relationship between the variables is linear. The points are sloping downwards: this means that as the selling price of milk goes up (increases), the weekly sales go down (decreases). This makes perfect sense. As the price of the product goes up, people tend to buy less or will look for a cheaper source of the product.

Figure 10.13: Scatter plot of weekly sales against selling price 1

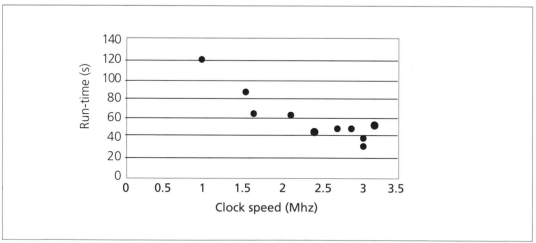

Activity 10.2: The relationship between computer speed and runtime

Plotting clock speed on the *x*-axis and run-time on the *y*-axis gives the plot in figure 10.14.

Figure 10.14: Scatter plot of run-time against clock speed

Once again, if we'd plotted the variables the other way round then the plot would look different but it wouldn't affect the problem.

From figure 10.14 there certainly seems to be a relationship between the points. The relationship does seem to be approximately linear, since we could draw a straight line which would go through or near all the points (although if you look more closely you can see that possibly the points lie closer to a curve, with the curve getting less steep for larger values of *x*). The points are sloping downwards, which means that as the clock speed increases, the run-time decreases. This is what we'd expect: computers with a faster clock speed take less time to run the test program.

Activity 10.3: Scatter plot of weekly sales against selling price 2

Figure 10.13 is the scatter plot associated with this data. The points in this scatter plot are not lying exactly on a straight line. Therefore this is partial correlation rather than perfect correlation. The points follow a suggested straight line quite closely so this is a reasonably strong relationship.

The same is true for the scatter plot in figure 10.14. The points lie close to a straight line but not exactly on it, so the relationship is strong but not perfect.

Activity 10.4: Scatter plot of weekly sales against selling price 3

Table 10.10 gives the details of the calculations for r. Variable x refers to the weekly sales and y to the selling price.

There are eight observations in each variable, so $n = 8$

$$\bar{x} = \frac{\sum x}{n} = \frac{88}{8} = 11 \qquad \text{and} \qquad \bar{y} = \frac{\sum y}{n} = \frac{272}{8} = 34$$

Table 10.10: Evaluating r

x	y	$x-\bar{x}$	$y-\bar{y}$	$(x-\bar{x})^2$	$(y-\bar{y})^2$	$(x-\bar{x}) \times (y-\bar{y})$
10	33	-1	-1	1	1	1
5	37	-6	3	36	9	-18
12	35	1	1	1	1	1
10	36	-1	2	1	4	-2
15	31	4	-3	16	9	-12
7	36	-4	2	16	4	-8
12	34	1	0	1	0	0
17	30	6	-4	36	16	-24
$\sum x$ = 88	$\sum y$ = 272			$\sum(x-\bar{x})^2$ = 108	$\sum(y-\bar{y})^2$ = 44	$\sum(x-\bar{x}) \times (y-\bar{y})$ = -62

Substituting the relevant sums into the formula

$$r = \frac{\sum(x-\bar{x})(y-\bar{y})}{\sqrt{\sum(x-\bar{x})^2 \times \sum(y-\bar{y})^2}}$$

gives

$$r = \frac{-62}{\sqrt{108 \times 44}} = \frac{-62}{\sqrt{4,752}} = \frac{-62}{\sqrt{68.93}} = 0.899$$

This correlation coefficient is negative as expected, so as the price of the milk goes up the amount sold goes down. This is not a perfect relationship as the correlation coefficient is not equal to -1. However, the correlation coefficient is quite close to -1, so the relationship is quite strong.

Activity 10.5: Relationship between computer speed and runtime 2

As in activity 10.2 we'll represent the clock speed by x the run-time by y. There are 10 pairs of observations, so $n = 10$. The means are

$$\bar{x} = \frac{\sum x}{n} = \frac{23.2}{10} = 2.32 \quad \text{and} \quad \bar{y} = \frac{\sum y}{n} = \frac{591}{10} = 5.91.$$

We can set the calculations out in a table as usual.

Table 10.11: Evaluating r

x	y	$x-\bar{x}$	$y-\bar{y}$	$(x-\bar{x})^2$	$(y-\bar{y})^2$	$(x-\bar{x}) \times (y-\bar{y})$
1.5	85	−0.82	25.9	0.6724	670.81	−21.24
3	39	0.68	−20.1	0.4624	404.01	−13.67
2.8	47	0.48	−12.1	0.2304	146.41	−5.81
2.1	63	−0.22	3.9	0.0484	15.21	−0.86
1.6	65	−0.72	5.9	0.5184	34.81	−4.25
2.4	43	0.08	−16.1	0.0064	259.21	−1.29
3.1	48	0.78	−11.1	0.6084	123.21	−8.66
1	118	−1.32	58.9	1.7424	3469.21	−77.75
2.7	48	0.38	−11.1	0.1444	123.21	−4.22
3	35	0.68	−24.1	0.4624	580.81	−16.39
1.5	85	−0.82	25.9	0.6724	670.81	−21.24
Σ 2.32	59.1			4.896	5826.9	−154.14

Substituting the sums into the formula we get

$$r = \frac{\sum(x-\bar{x})(y-\bar{y})}{\sqrt{\sum(x-\bar{x})^2 \times \sum(y-\bar{y})^2}} = \frac{-154.14}{\sqrt{4.896 \times 5826.9}} = \frac{-154.14}{\sqrt{28528.50}} = \frac{-154.14}{168.90} = -0.912$$

This correlation coefficient is negative as expected so, as the clock speed increases, the run-time decreases. The correlation coefficient is close to −1 (even closer than in activity 10.4), so the relationship is strong but not perfect.

Activity 10.6: Does watching television extend your life?

This is clearly an example of spurious correlation. Both variables are affected by a third variable, which is the wealth of the country. Richer countries probably have more televisions than poorer countries. Richer countries also tend to have longer life expectancies because they have better nutrition, cleaner water, better health facilities etc. There is no cause and effect link between owning televisions and length of life.

Activity 10.7: Do policemen affect the price of chocolate?

This is an example of spurious correlation, although in this case there is probably an indirect link between the variables. The direct links suggested in the question aren't real, but more general versions of them are. *Overall*, workers are paid with the earnings of companies, and *overall* if wages rise then workers have more money to spend, so prices rise. Thus both policemen's wages and the price of chocolate are linked to the level of inflation: if inflation is high, then wages and prices both rise, giving a positive correlation.

Activity 10.8: Do bananas cure polio?

This is certainly an example of spurious correlation. The reason there seems to be a link is that two events happened at about the same time in the mid-twentieth century: it became economic to import bananas into the UK, and an effective polio vaccine was developed and began to be widely used.

As with activity 10.6 we might be able to tell that something is wrong by plotting both variables against time. We should find that the number of bananas eaten suddenly begins to rise at some point, and the number of polio cases suddenly begins to fall at some point. Even if these happen at about the same time, it suggests that there may be other factors at work on each variable, rather than there necessarily being a direct link between the two variables.